LAND-USE PLANNING FOR SUSTAINABLE DEVELOPMENT

SECOND EDITION

Social-Environmental Sustainability Series

Series Editor

Chris Maser

Published Titles

Land-Use Planning for Sustainable Development, Second Edition
Jane Silberstein and Chris Maser

Insects and Sustainability of Ecosystem Services
Timothy D. Schowalter

Biosequestration and Ecological Diversity: Mitigating and Adapting to Climate Change and Environmental Degradation
Wayne A. White

Decision-Making for a Sustainable Environment: A Systemic Approach
Chris Maser

Economics and Ecology: United for a Sustainable World
Russ Beaton and Chris Maser

Resolving Environmental Conflicts, Second Edition
Chris Maser and Carol A. Pollio

Fundamentals of Practical Environmentalism
Mark B. Weldon

Sustainable Development: Principles, Frameworks, and Case Studies
Okechukwu Ukaga, Chris Maser, and Michael Reichenbach

Social-Environmental Planning: The Design Interface Between Everyforest and Everycity
Chris Maser

Social-Environmental Sustainability Series

LAND-USE PLANNING FOR SUSTAINABLE DEVELOPMENT

SECOND EDITION

JANE SILBERSTEIN

CHRIS MASER

CRC Press
Taylor & Francis Group
Boca Raton London New York

CRC Press is an imprint of the
Taylor & Francis Group, an **informa** business

Cover image: Ros Nelson (Little Big Bay, LLC)

CRC Press
Taylor & Francis Group
6000 Broken Sound Parkway NW, Suite 300
Boca Raton, FL 33487-2742

First issued in paperback 2019

ISBN-13: 978-1-4665-8114-2 (hbk)
ISBN-13: 978-0-367-86804-8 (pbk)

Library of Congress Cataloging-in-Publication Data

Silberstein, Jane.
Land-use planning for sustainable development / Jane Silberstein and Chris Maser.
-- Second edition.
pages cm
Includes bibliographical references and index.
ISBN 978-1-4665-8114-2
1. Land use--United States--Planning. 2. Sustainable development--United States. I. Maser, Chris. II. Title.

HD205.S55 2013
333.73--dc23 2013018312

Visit the Taylor & Francis Web site at
http://www.taylorandfrancis.com

and the CRC Press Web site at
http://www.crcpress.com

Contents

Editor's Note

The book you are holding is part of a series on the various aspects of social-environmental sustainability. *Land-Use Planning for Sustainable Development* focuses on the primacy and quality of relationships among people sharing a particular place and between those people and their environment. "Development" means personal and social transformation to a higher level of consciousness and a greater responsibility toward the next generation. "Sustainability" is the act whereby one generation saves options by passing them to the next generation, which saves options by passing them to the next, and so on.

This series of books on the various facets of social-environmental sustainability is a forum in which those who dare to seek harmony and wholeness can struggle to integrate disciplines and balance the material world with the spiritual, the scientific with the social, and in so doing expose their vulnerabilities, human frailties, and hope, as well as their visions for a sustainable future.

In writing this book, we are reminded of a comment author Scott Nearing noted many years ago when he wrote on a small card, "The majority will always be for caution, hesitation, and the status quo—always against creation and innovation. The innovator—he [or she] who leaves the beaten track—must therefore always be a minoritarian—always be an object of opposition, scorn, hatred. It is part of the price he [or she] must pay for the ecstasy that accompanies creative thinking and acting."

As the title of this book implies, *Land-Use Planning for Sustainable Development* is part of our human journey toward the ideal of social-environmental sustainability as an unconditional gift from the present generation to those of the future. Although some people are quick to point out that ideas, such as those expressed in this book, are against what society has come to unquestioningly accept as "human nature," we disagree. This notion is unacceptable when our present course is inextricably impoverishing each successive generation. Besides, those who are afraid of change inevitably point to ideas that differ from their own and say they are impractical. However, so-called "impracticality" is merely a horizon of ideas that have not yet been tested. Until they are, how does one know they are "impractical?"

Chris Maser, Series Editor

Preface

> Whether we and our politicians know it or not, Nature is party to all our deals and decisions, and she has more votes, a longer memory, and a sterner sense of justice than we do.
>
> **—Wendell Berry**[1]

The first edition of *Land Use Planning for Sustainable Development* was published in 2000. Since then, the world has reeled from illusion-shattering events like 9/11, the Wall Street bailout, new U.S.-led wars, and a prediction that by 2035 the Arctic ice cap will disappear.[2] Over this same period, U.S. citizens have suffered alarming infringements on constitutional rights, surging income inequality, and for many, loss of life or the life-changing impacts of megastorms like Sandy and Katrina. Yet renewal, *along with collapse,* continues, and the impartial forces of Nature persist. The life-death cycle remains a fixture. And, while the Mayan prophecy, which predicted the end of the world in December 2012, was incorrect, many still feel it is about to happen. Our hope is this book will help reframe this perspective by showcasing creative, resilient strategies for fundamentally changing the way we alter our landscape.

Change is a constant process of creating novelty and thus new realities. These new realities served to inform and inspire the reshaping of this book. We searched for examples of innovation that have emerged from and responded to the disruption and trauma of the past 13 years. We discovered the phoenix in the writings of visionary authors, in the accomplishments of grassroots organizations, and in the bold strokes and successes of self-directed communities.

Since all dimensions of land use are intimately related to economics, property, and ownership, we turned to the most thoughtful as well as hopeful new perspectives, including the works of David Korten, who envisions a new economy in *Agenda for a New Economy* (first and second editions)[3]; Russ Beaton and Chris Maser, who envision a sustainable, social-environmental economic system in *Economics and Ecology: United for a Sustainable World,*[4] based on an understanding and acceptance of nature's inviolable biophysical principles as reflected in the operative principles of economics; and the alternative ownership models proposed by Marjorie Kelly in *Owning Our Future: The Emerging Ownership Revolution.*[5]

We also document innovative approaches to development in small communities, like Bayfield, Wisconsin, as well as larger urban centers like Salt Lake City, Santa Monica, and Seattle, and progressive citizen-led efforts, such as the Living Future Institute, whose mission is to lead and support the transformation toward communities that are socially just, culturally rich, and ecologically repairable. Similarly, the Business Alliance for Local Living Economies envisions "within a generation … a global system of human-scale, interconnected local economies that function in harmony with local ecosystems to meet the basic needs of all people, support just and democratic societies, and foster joyful community life."[6] We revisit the successes as well as barriers to progress associated with establishing new community development models, such as ecomunicipalities.

In this edition, in four new chapters, we attempt to better frame our message regarding the need for reinventing land use practices. We have also made significant changes through reorganization of the material from the first text, as well as updated all chapters from the first edition with current examples and pertinent references.

What we are most excited about is our proposal for linking land use planning and regulation with nature's inviolate biophysical principles. While a work in progress, we believe we offer a strong foundation and inspiration for the ongoing development of new approaches.

Most important, this new edition emphasizes the necessity and potency of citizen involvement and initiatives. As we all know at this point in history, adaptation is critical, and, as Vicki Arroyo, executive director of the Georgetown Climate Center, stated in a TED (Technology, Entertainment, Design) talk in 2012, "Adaptation is too important to be left to the experts."[7] The social, economic, and environmental challenges facing society today are daunting and complex. Solving them will demand a systemic worldview and an ability to move beyond the level of the problem, where awareness is limited, to the level of the solution, where we consciously look at beliefs, assumptions, expectations, perceptions, and even emotions that may be limiting our ability to find solutions.

In a sea of unknowns and complexity, maintaining a focus on the health and functioning of our relationships with one another and the planet is critical. Further, such a focus is essential for creating a land use planning paradigm that (1) embraces the complex needs, interrelationships, and interdependencies of humans and other life forms; (2) incorporates ecological principles into the process and results; and (3) converts the planning process to a cycle in which the "plan" is only one part of an ongoing process to the benefit of all generations—present and future.

As you read this book, bear in mind that our ideas and proposals for an alternative approach to land use rest on lessons from history, as well as the research, wisdom, and vision of many individuals and communities whose work we have studied. With the imperative of constantly adapting to changing conditions, our hope is that what we propose proves to be a sturdy platform on which to continually build and innovate.

> All truly wise thoughts have been thought already thousands of times, but to make them truly ours, we must think them over again honestly, til they take root in our personal experience.
>
> **—Johann Wolfgang von Goethe**[8]

Endnotes

1. Wendell Berry. http://www.goodreads.com/quotes/758-whether-we-and-our-politicians-know-it-or-not-nature (accessed November 5, 2012).
2. National Oceanic and Atmospheric Administration. "Sea Ice." n.d. http://www.arctic.noaa.gov/future/sea_ice.html (accessed November 5, 2012).
3. (1) David Korten. *Agenda for a New Economy: From Phantom Wealth to Real Wealth.* Berrett-Kohler, San Francisco, 2010; and (2) David Korten. *Agenda for a New Economy: From Phantom Wealth to Real Wealth,* 2nd edition. Berrett-Kohler, San Francisco, 2012.
4. Russ Beaton and Chris Maser. *Economics and Ecology: United for a Sustainable World.* CRC Press, Boca Raton, FL, 2012.
5. Marjorie Kelly. *Owning Our Future: The Emerging Ownership Revolution.* Berrett-Kohler, San Francisco, 2012.
6. Business Alliance for Local Living Economies. "About BALLE." http://bealocalist.org/about-us (accessed November 5, 2012).
7. Georgetown Climate Center. "Resources for Adaptation Action." n.d. http://www.georgetownclimate.org/node/4025 (accessed November 5, 2012).
8. Johann Wolfgang von Goethe. http://www.quotationspage.com/quote/9806.html (accessed November 5, 2012).

1

Sustainable Development: The Concept

Introduction

An awareness of disastrous consequences, brought about by historically proven unwise choices, provides us with the opportunity to change the way we do things so that we can alter an outcome. But, before we can alter the outcome of any historical trend, we must ask fundamentally different questions.

One such question is: What is sustainable community development? From our perspective, it is a process of continuous improvement in which the community is constantly evolving and making changes to accomplish its goals. *Continuous improvement* in this context means a nonlinear process of systems thinking that accounts for the social significance of nonmaterial wealth, qualitative values, and the heritage of both cultural diversity and identity. It is a process that views humans as an inseparable part of nature and development as directed change leading to social maturation. The resulting psychological transformation is characterized, in part, by people voluntarily reaching beyond the immediacy of self-centeredness in a conscious effort to be accountable for the effects present decisions may cause in the future. Development thus becomes a process of change guided by the principles of social-environmental justice for all living things present and future, not just humans.

To understand this better, we have identified eleven essential elements of sustainability:

1. Understanding and accepting that we cannot touch any part of a system without affecting the whole of it—the "water bed principle";
2. Understanding and accepting nature's inviolate biophysical principles, which determine the quality of human survivorship;
3. Understanding and accepting that we do not—and cannot—manage nature;
4. Understanding and accepting that we tend to make an ecosystem more fragile when we alter it;
5. Understanding and accepting that we must reinvest in living systems even as we reinvest in businesses;

6. Understanding and accepting that only a systemic worldview is a sustainable worldview;
7. Accepting our ignorance and trusting our intuition while doubting the certainty of our knowledge;
8. Specifying what is to be sustained;
9. Understanding and accepting that sustainability is a continual process, not a fixed end point;
10. Understanding, accepting, and being accountable for saving choices and things of value from which to choose and passing them forward to the next generation;
11. Understanding, accepting, and being accountable for biophysical limitations to land "ownership" and the rights of "private property."

The Elements of Sustainability

To achieve the balance of energy necessary to sustain vital, healthy ecosystems, we must focus our questions, both social and scientific, on an understanding of the inviolable biophysical principles that govern those systems. Then, we must find the moral courage and political will to direct our personal and collective energy toward living within the constraints defined by these principles.

Adapting our lives and our processes to fit within these limits is the goal, all the while factoring in the inseparability of our human presence. Everything we do alters our environment, and the systems we create take us further from the biophysical balance that characterized the prehuman condition. And, since change is a continual, irreversible process, we must keep in mind that *restoration* of a specific condition is not possible, in spite of its broad use as a label for projects aimed at environmental repair.

In addition, we must keep in mind that sustainability is not a condition in which compromise can be struck, although decisions leading toward sustainability often require compromise. Seeking sustainability to a degree, which appears like an innocuous compromise, defeats sustainability altogether. Leave one process out of the equation or in some other way alter a feedback loop, and the system as a whole will gradually be deflected toward an outcome other than that which was originally intended. It is thus critical during any planning process to consider carefully the elements of sustainability, understand and accept their inviolability, and thus remain within the biophysical limitations that make them work, if a given land use plan is to be viable through time.

The following nonnegotiable elements of sustainability cover only those that we have found to be common impediments to sustainable land use planning.

First Element: The Water Bed Principle

We are an inseparable part of nature, and just as with a water bed, touching any part affects the whole. Air, for example, is the interactive thread connecting soils, water, biodiversity, human population density, sunlight, and climate. (Biodiversity refers to the variety of living species and their biophysical functions and processes.) Our earthly survival, and that of our children and their children unto all generations, depends on clean air.

But air, like other natural resources, has long been considered by most modern societies to be unlimited and free for the taking. "The history of almost every civilization," observed British historian Arnold Toynbee, "furnishes examples of geographical expansion coinciding with deterioration in [environmental] quality."[1] Fortunately, today, there is more talk about "renewable" resources and much less talk about unlimited resources.

Ultimately, however, all biophysical resources are finite—with the exception of solar energy, at least for the purposes of this book. We not only can run out of a resource by literally exhausting its earthly supply, such as oil or the extinction of a species and its attendant ecological service, but also can alter an existing resource to render it useless to us, such as poisoning our drinking water through pollution of various kinds. And, we are increasingly doing both.

Moreover, as the burgeoning human population demands more and more material commodities from a rapidly dwindling supply of an increasing number of vital necessities (such as potable water), the ratio of resources apportioned to each human declines. This decline is exacerbated by the progressively longer lives of today's humans and their prolonged demands for these same necessities. Further, those resources currently deemed "renewable" are only renewable as long as the system producing them retains its biophysical integrity and is used in a sustainable manner—both ecologically and socially, as clearly stated by T. N. Narasimhan of the Berkeley Institute of the Environment:

> Historically, science has pursued a premise that Nature can be understood fully, its future predicted precisely, and its behavior controlled at will. However, emerging knowledge indicates that the nature of Earth and biological systems transcends the limits of science, questioning the premise of knowing, prediction, and control. This knowledge has led to the recognition that, for civilized human survival, technological society has to adapt to the constraints of these systems. Simultaneously, spurred by explosive developments in the understanding of materials (non-biological and biological), applied scientific research pursues a contrary goal of controlling the material world, with the promise of spectacular economic growth and human well-being. If adaptation to Nature is so important, why does applied research pursue a contrary course? ... Also, in a world dominated by democratic ideals of freedom and liberty, the discipline required for adapting to Nature may often be overridden by competition among various segments of society to exercise their respective rights.[2]

Part of our inability to intellectually grasp the eternal mystery of our earthly relationships is that no two things in the universe have ever been—or ever will be—exactly the same. Therefore, no two things can ever be equal—except in their inequality. Moreover, all life is composed of physical relationships in ever-changing patterns and rhythms that both affect life and are, in turn, affected by life. In this sense, life not only is pattern-seeking and pattern-sensitive but also is guided by the eternal rhythms of the universe. As well, every life form is a microcosm of the whole—from the simplest to the most complex.

Everything in the universe is thus connected to everything else in a cosmic web of interactive feedback loops, all entrained in self-reinforcing relationships that continually create novel, never-ending stories of cause and effect. Everything, from a microbe to a galaxy, is defined by its ever-shifting relationship to every other component of the cosmos. Thus, "freedom" (perceived as the lack of constraints) is merely a continuum of fluid relativity. Hence, every change (no matter how minute or how grand) constitutes a systemic modification that produces novel outcomes. A feedback loop, in this sense, comprises a reciprocal relationship among countless bursts of energy moving through specific strands in the cosmic web that cause forever-new, compounding changes at either end of the strand, as well as every connecting strand.

Relationships among things are in constant flux as complex systems arise from subatomic and atomic particles in the giant process of evolution on Earth. At each higher level of complexity and organization, there is an increase in the size of the system and a corresponding decrease in the energies holding it together. Put differently, the forces that keep evolving systems intact, from a molecule to a human society, weaken as the size of the systems increases, yet the larger the system the more energy it requires to function. Such functional dynamics are characterized by their diversity as well as by the constraints of the overarching laws and subordinate principles that govern them.

Second Element: Understanding and Accepting Nature's Inviolate Biophysical Principles

These principles can be said to govern the world and our place in it because they form the behavioral constraints without which nothing could function in an orderly manner. In this sense, the law of cosmic unification is analogous to the Constitution of the United States, a central covenant that informs the subservient courts of each state about the acceptability of its governing laws. In this way, nature's rules of engagement inform society of the latitude whereby it can interpret the biophysical principles and survive in a sustainable manner. Natural systems will always function in a self-sustaining manner but not necessarily in a way that is favorable to human survival, let alone a life of dignity.

The law of cosmic unification is functionally derived from the synergistic effect of three universal laws: the first law of thermodynamics, the second law of thermodynamics, and the law of maximum entropy.

The *first law of thermodynamics* states that the total amount of energy in the universe is constant, although it can be converted from one form to another. Therefore, the amount of energy remains the same, even if you could go forward or backward in time. For this reason, the contemporary notions of "energy production" and "energy consumption" are oxymorons. The *second law of thermodynamics* states that the amount of energy in forms available to do useful work can only diminish over time. The loss of available energy to perform certain tasks thus represents a diminishing capacity to maintain order at a certain level of manifestation (for example, a tree) and so increases disorder or entropy. This "disorder" ultimately represents the continuum of change and novelty—the manifestation of a different, simpler configuration of order, such as the remaining ashes from the tree when it is burned. In turn, the *law of maximum entropy* says that a system will select the path from available paths that maximizes entropy at the fastest rate given the existing constraints.[3]

The essence of maximum entropy simply means that when any kind of constraint is removed, the flow of energy from a complex form to a simpler form speeds up to the maximum allowed by the relaxed constraint.[4] We, in contemporary society, are all familiar with the basis of this law even if we do not understand it. For example, when our body loses heat in cold weather, our sense of heat loss increases exponentially when wind chill is factored in because our clothing has ceased to be as effective a constraint to the loss of heat as it was before the wind became an issue. Moreover, the stronger and colder the wind, the faster our body loses its heat—the maximum entropy (= loss) of our body's energy whereby we stay warm. If the loss of body heat to wind chill is not constrained, hypothermia and death ensue, along with the beginnings of bodily decomposition—reorganization from the complex structure and function toward a simpler structure and function.

With respect to the functional melding of these three laws, Rod Swenson of the Center for the Ecological Study of Perception and Action, Department of Psychology, University of Connecticut, says these three laws of thermodynamics "are special laws that sit above the other laws of physics as laws about laws or laws on which the other laws depend" (p. 315).[5] Stated differently, these three laws of physics coalesce to form the *supreme* "law of cosmic unification," to which *all* biophysical and social principles governing nature and human behavior—decision making—are subordinate yet simultaneously inviolable. *Inviolable* means that we manipulate the effects of a principle through our decisions and subsequent actions, but we do not—and cannot—alter the principle itself. The biophysical principles, which are discussed in detail in Chapter 3, are:

- Principle 1: Everything is a relationship.
- Principle 2: All relationships are inclusive and productive of an outcome.

- Principle 3: The only true investment in our global ecosystem is energy from sunlight.
- Principle 4: All systems are defined by their function.
- Principle 5: All relationships result in a transfer of energy.
- Principle 6: All relationships are self-reinforcing feedback loops.
- Principle 7: All relationships have one or more trade-offs.
- Principle 8: Change is a process of eternal becoming.
- Principle 9: All relationships are irreversible.
- Principle 10: All systems are based on composition, structure, and function.
- Principle 11: All systems have cumulative effects, lag periods, and thresholds.
- Principle 12: All systems are cyclical, but none is a perfect circle.
- Principle 13: Systemic change is based on self-organized criticality.
- Principle 14: Dynamic disequilibrium rules all systems.

Third Element: Understanding and Accepting That We Do Not and Cannot Manage Nature

Sustainability is predicated on the humility to admit and accept that we humans do not—and cannot—manage nature, that we are not in control. Rather, we treat nature in some way, nature responds to that treatment, and a feedback loop is set in motion.

Here, T. N. Narasimhan asks a pertinent question:

> How then may humans sustain on a finite earth in which behavioral patterns of life-supporting natural resource systems can be understood, but cannot be predicted or controlled? Reason suggests that society must adapt its functioning to the constraints imposed by the nature of these resource systems. This general need to structure patterns of living within constraints that lie beyond human understanding or control may be referred to as "adapting to Nature."[6]

Our human insistence on managing what cannot be managed (a tree farm or a population of salmon or tuna) while leaving *unattended* what could be *cared for* sustainably (a forest or an ocean) is one source of the problem. In addition, our lack of will to control our materialistic appetites, expansive economies, exploitive technologies, and exploding population is causing us to unintentionally redesign Earth. That we redesign Earth is a given simply because we exist and use energy, but how we go about redesigning Earth is another matter.

The paradigm within which our social decisions are generally made today in economics, agriculture, ocean fisheries, and the determination of yields in forestry is based on the illusion of limitless supply. This economic view of quantitatively constant values, such as independent variables, such as the notion of the unchanging biological fertility of the soil, stability of climate, and dismissal of both biological and genetic diversity as unnecessary puts us at peril since this view ignores the reality of time and change.

This is revealed in such concepts of a sustained yield in timber, sustained yield in finances, and sustained growth in economics. Evolution, in this view, is reduced to a mechanical process without any place for the novelty of change. It is a clockwork world with no capacity for a creative process. It theoretically allows only reversible locomotion devoid of creativity, like a train running endlessly back and forth on a track.

The consequences of this perspective for traditional economics is that resources—soil, water, air, sunlight, biological diversity, genetic diversity, and climate—are no longer considered "free," and their degradation thus comes with "costs." Traditional economics' view of resources as free means resources belong to no one and thus to everyone, so their quick exploitation makes sense; otherwise, their perceived short-term economic value would be lost either to someone else or to nature through an ecological "catastrophe," such as a forest fire.

Thus it is that the timber industry, having nothing invested in the growing of the old-growth trees, sees the ancient forest as free and all too often ignores the degradation of the land caused by logging. The timber industry contends that the wood fiber it harvests is a product of the air and sunlight and thus ignores the qualitative depreciation of the biological capital of the soil, which is largely governed by biological diversity, genetic diversity, functional diversity, and climate.

The timber industry has also tended to ignore the creative evolution of the forests it harvests, evolution that is ongoing in the genetic, biophysical, and cultural sense. Norman Jacob puts it this way: "The 'normal' forests of industrialists and the 'perfect' markets of economists are timeless, qualitiless, mechanical, and strictly fictional constructs of what in reality is a timeful, qualitative, and organic world" (p. 95).[7]

In such a static world, devoid of change and without novelty, efficiency of use in resources is an attractive concept because it is easy to *know* what is best in a predictable, objective, static, and mechanical world. In such a world, the perspective of relationship in space and the irreversible quality of time do not exist. The economic goal thus becomes the homogenization of the creative, evolutionary process through so-called objectivity, whereas human subjectivity—the reality of our perspectives—is summarily invalidated.

When we insist on optimizing or doing what we *intellectually* "know is best" biophysically, it is inevitably from an economic/political point of view, which is almost always counterintuitive as well as socially and biophysically unwise. If, however, we honor our intuition, we will find the place where culture meets nature and both are more nearly sustainable.

Fourth Element: Understanding and Accepting That We Make an Ecosystem More Fragile When We Alter It

Ecosystems, *which include all human social systems,* are designed by the *variability of the variability* of natural phenomena, such as volcanoes, climate, fires, floods, and the cyclical nature of populations of organisms. They are not designed by the predictable averages of anything. Ecosystems are designed by novelty and uncertainty, not by static certainty.

In addition, while incremental changes in an ecosystem may seem to be insignificant and their effects for a time to be invisible, ecosystems operate on thresholds with unknown margins of safety. We humans, on the other hand, live in the "invisible present" and thus cannot always recognize cumulative effects.[8]

The invisible present is our inability to stand at a given point in time and see the small, seemingly innocuous effects of our actions as they accumulate over weeks, months, and years. Obviously, we can all sense change—day becoming night, night turning into day, a hot summer changing into a cold winter, and so on—but we typically underestimate the degree to which slow, seemingly innocuous change has occurred, such as global warming. We are unable to sense slow changes directly, and we are even more limited in our abilities to interpret the relationships of cause and effect in these changes. Nevertheless, once a threshold is crossed—it is crossed. There is no going back to the original condition because we cannot go backward in time. It is thus necessary to understand something about the relative fragility of simplified ecosystems as opposed to the robustness of complex ones.

Fragile ecosystems can go awry in more ways and can break down more suddenly and with less warning than is likely in robust ecosystems because fragile systems have a larger number of components with narrow tolerances than do robust ones. As such, the failure of any component can disrupt the system. Therefore, when a pristine ecosystem is altered for human benefit, it is made more fragile, which means that it will require more planning and maintenance to approach the stability of its original condition. Thus, while sustainability means maintaining the critical functions performed by the primeval system, or some facsimile thereof, it does not mean restoring or maintaining the primeval condition itself, which is a biophysical impossibility.

If one thus looks at ecosystems along a continuum of naturalness (the most pristine being the most natural end of the continuum and the most human-altered being the most cultural end of the continuum), the notion of system fragility not only makes sense but also offers humanity a range of choices. And it is, after all, the array of choices one generation passes to the next that conveys the sustainability of potential outcomes.

For example, the less we humans alter a system to meet our necessities, the more the system's functional requirements are met internally. This in turn makes it easier and less expensive in both time and energy (including money) to maintain that system in a *relatively* steady state because we have maintained more diversity within the native flora and fauna and thus buffered the functionality of the biophysical processes from human intrusions.

Conversely, the more altered a system is, the more that system's functional requirements must be met through human-mediated sources external to itself. This in turn makes it more labor intensive and more expensive to keep that system in a given condition because we have maintained less, often far less, of the diversity of its indigenous flora and fauna.

But why, we might ask, do we need such a variety of species? What effect does a variety of species have on the health of an ecosystem? One marvelous effect such variety has is increasing the stability of ecosystems through feedback loops, the means by which processes reinforce themselves.

Strong, self-reinforcing feedback loops characterize many interactions in nature and have long been thought to account for the stability of complex systems. Ecosystems with strong interactions among components, reinforced by feedback loops, can be complex, productive, relatively stable, resilient, and thus adaptable to changing conditions. When, however, these critical loops are disrupted, as in the extinction of species and the loss of their biological functions, these same systems become more fragile and are easily affected by slight changes.

It is the variety of species that creates the feedback loops. Each species, by its very existence, has a shape and therefore a structure, which in turn allows certain functions to take place, functions that interact with those of other species. That is what makes each individual species so valuable. All of this is governed ultimately by the genetic code, which by replicating species' character traits builds critical "backups" into each ecosystem, which protect its functional abilities.

A *backup* means that more than one species can perform similar functions. It is a type of ecological insurance policy that strengthens the ability of the system to retain the integrity of its basic relationships. The insurance of a backup means that the loss of a species or two is not likely to result in ecosystem collapse—in terms of its ability to serve our human requirements—because other species can make up for the functional loss. But, there comes a point, a threshold, however, when the loss of one or two additional species may in fact tip the balance and cause the system to begin an irreversible decline in terms of *human-related* productivity or value.

Although an ecosystem may be stable and able to respond "positively" to the disturbances in its own environment to which it is adapted, this same system may be exceedingly vulnerable to the introduction of foreign disturbances (often those introduced by humans) to which it is not adapted. We can avoid disrupting an ecosystem supported by feedback loops only if we understand and protect the critical interactions that bind the various parts of the system into a functional whole.

Diversity of plants and animals therefore plays a seminal role in buffering an ecosystem against disturbances from which it cannot recover. As we lose species, we lose not only their diversity of structure and function but also their genetic diversity, which sooner or later results in complex ecosystems becoming so simplified they will be unable to sustain us as a society. Therefore, any societal strategy aimed at protecting diversity and its evolution is a critically important step toward ensuring an ecosystem's ability to adapt to change. Diversity counts. We need to protect it at any cost.

Although ecosystems can tolerate cultural alterations, those functions that have been disrupted or removed in the process (often through a loss of species) must be replaced through human labor if the system is to be sustainable. The more a system is altered and simplified, the more fragile it becomes and the more labor intensive its maintenance becomes. When cultural manipulation exceeds the point at which human labor can maintain the necessary functions, the system is said to "collapse," which really means that it can no longer serve our social requirements.[9]

Collapse is not always caused by dynamics internal to the system itself, however. To understand this, we need to look at one ecosystem (Easter Island) that does appear to have collapsed from within as a basis of comparison for others that did not.

Self-Destruction

The story of Easter Island is a presage of increasing self-destruction in the world if we do not control our human population. And, the only way for such control to take place is to grant women unequivocal equality in all aspects of life. Should you dismiss human overpopulation as a global problem while reading this brief history, keep in mind that overpopulation is only a matter of scale within any ecosystem across a landscape.

Easter Island is a tiny, 43-square-mile (111-square-kilometer) piece of land 2,400 miles (3,862 kilometers) off the coast of South America in the South Pacific. The oldest pollen dates (analysis of the pollen from certain ancient plants) go back some 30,000 years, long before the first people, wandering Polynesians, arrived. At that time, based on the pollen record, the island was forested with now-extinct, giant Jubaea palms.

The Polynesians settled on the island in about 1200 AD. They began to gradually clear the land for agriculture, and they cut trees to build canoes. The island, while small, was relatively fertile, the sea teemed with fish, and the people flourished. The population rose to 3,000 or 4,000 and probably remained relatively stable, given the low carrying capacity of the island. Eventually, however, trees were cut to provide logs for transporting and erecting hundreds of statues, or *moai,* some of which are about 32 feet (9.8 meters) high and weigh as much as 85 tons (77 metric tons).

Unfortunately, when the trees were cut, they did not grow back, as shown in the pollen record. Deforestation, which began shortly after the first people arrived in 1200, was almost complete 500 years later, by 1700. When the Europeans discovered Easter Island in 1722, it was treeless and in a state of decline; nevertheless, the Dutch explorer Jacob Roggeveen and the commanders of his three ships described the island as "exceedingly fruitful, producing bananas, potatoes, sugar-cane of remarkable thickness, and many other kinds of fruits of the earth." If the soil was rich enough for these plants, why did the trees not grow back?

As it turns out, the Polynesians brought the Polynesian rat with them in their boats. Once the rats were on the island, they discovered and began to devour the palm nuts. So, while the human population was expanding and the people were cutting the trees, the rat population was also expanding, eating the palm nuts, and thus preventing the trees from growing back. As well, the effects of drought, wind, and soil erosion could have exacerbated the island's deforestation. In addition, both the people *and* the rats exploited many of the island's other resources, such as its abundance of birds' eggs. The downward spiral had begun.

With deforestation, there were no trees to build canoes for fishing. Soil erosion would lead to smaller-than-historic crop yields. And, the eggs of the sooty tern were probably exploited to the point that the bird no longer nested on the island.

Fewer fish, eggs, and crops led to a shortage of food. Hunger, in turn, eventually brought the civilization to the brink of collapse. Today, all that remains of the original culture of Easter Island are the statues that once stood erect on specially built platforms, others that lie abandoned between the volcanic quarries of their origin and their planned destinations, and still others that remain unfinished in the quarries.[10]

Loss of Labor Pool

In contrast, consider ancient Greece, which flourished under wise agricultural use during the beginning of the Iron Age. Nevertheless, the Greeks had greatly altered the landscape, despite their apparently sound agricultural ethic. But, all the human-caused changes, including deforestation, do not appear to have caused the collapse of the agricultural system. In fact, to the best of our knowledge, it not only was sustainable and was being sustained but also might have continued to the present day if it not been for the effect of outside influences.

While the Greeks modified their landscape, making it fragile, their agricultural system was sustainable as long as there was a full human population to tend to the terraced fields. The destruction of their agricultural system was not a consequence of the system itself, but rather due to loss of the labor pool caused by Romans raiding the Greek countryside for slaves, a practice that reduced the population and left the fragile landscape untended to wash into the sea. Thus, as long as the Greeks maintained adequate cover crops, which were both labor intensive and functioned to hold the soil in place as the forests had once done, their agricultural system was sustainable. But, as the activities of Roman slavers continually reduced the Greek population, there came a threshold beyond which this labor-intensive agriculture simply could not be maintained and therefore collapsed.

Prior to the advent of Greek agriculture, the land had been forested for millennia, making sustainability a moot point. Sustainability arose as a problem not because of deforestation but rather because of the inability of a society debilitated by the slave trade to continue performing the function of the forest, namely, conservation of the soil.[11]

In fact, the emission of pollutants has tripled the level of nitrogen some forests are receiving.[12] Change in the level of nitrogen works as follows: People are used to the idea that if one puts nitrogen on a corn field, the corn grows faster, but then corn happens to be a plant that opens its stomata (pore-like entrances into the leaves) and takes up more carbon to create a balance. Thus, up to some point, the more nitrogen corn is given, the more carbon it will acquire, and the better it will grow. Unlike corn, however, most forest species will not grow any faster with additional nitrogen.

Introduced Technology

In addition to a physical loss of individuals from a communal labor pool, introduced technology, which replaces traditional human labor without removing the people from the potential labor pool, can also cause an internally sustainable social-environmental system to collapse. In June 1992, I (Chris) visited Čergov, in northern Slovakia, to evaluate the condition of the native forest, which is primarily European beech with an admixture of white fir. The native forest was being rapidly clear-cut and replaced with plantations of such nonnative species as Norway spruce, larch, and pine. The biological errors of forestry made in Germany, the United States, and Canada (replacing biologically diverse forests with short-lived, simplistic, economic tree farms) were being repeated in the forests of Čergov and for the same reasons—immediate and short-term economic gain.

Among the most graphic examples of technology disrupting a centuries-old, sustainable system of selective logging was the recent clear-cutting and its resultant loss of topsoil. Within 1 hour after each thunderstorm in Čergov, all the streams and rivers fed by clear-cut slopes went from clear water to the appearance of milk chocolate as the soil of the forest was washed away to the sea.

Prior to importing the technology of clear-cutting, the forest of Čergov was selectively logged with horses. In addition, horse logging had been biologically sustainable for centuries, as were the economies of the small mountain villages located in the upper valleys near the edge of the forest. Now, the jobs once sustained by horse logging are gone, like the topsoil of the forest. The people who once made their living from the forest must commute to the cities to find work, and the villages have lost part of their cultural heritage.

Once outside desires, such as financial gain, become intertwined in the balance of an internally sustainable social-environmental system, it is easy to ignore, and thus lose sight of, a sound ecological perspective. For example, a forester from a distant city told me that it was inevitable the old trees of Čergov would be clear-cut, and that plantations of spruce would replace them. I questioned this statement because there was nothing inevitable about either clear-cutting the old trees or planting spruce in rows. It was simply someone's choice to maximize immediate economic gain.

The forester told me that clear-cutting was necessary because the ground was too steep for logging with horses. I found this statement to be particularly interesting because the forest had been logged with horses for centuries before the chain saw and log truck became available, and all the clear-cuts I saw could easily have been logged with horses. Although such technology as chain saws and log trucks has the ability to directly disrupt local ecosystems, the effects of technology can grossly alter ecosystems at great distances from the technological apparatus itself.

Long-Distance Transport of Air Pollutants

We pollute the air with chemicals. Air pollution directly affects vegetation by altering the quality of the soil and water as well as the quality and quantity of the sunlight that drives the plant/soil processes. The chemicals we dump into the air also alter the climate and thus the environment in which the vegetation grows.

To illustrate, consider acid rain, which has long been recognized as a pollution problem in Europe, where statues and gargoyles that once adorned city streets and plazas and guarded centenarian buildings have had their faces dissolved over recent decades. The statues that I (Chris) remember seeing as a boy, in perfect form and feature, today are often-unrecognizable relics of a past era because acid rain has eaten away the marble much as leprosy eats away the flesh.

Acid rain is not confined to European cities, however. It is also found in forest and fen, in highland and lowland. There also it is destroying the essence of life as it joins league with other forms of industrial/technological pollution, where it contributes to a phenomenon the Germans call *Waldsterben*, the dying forest.

The dying forest syndrome is not exclusively the property of Europe. Every industrial country, including the United States and Canada, owns it. Here, called forest dieback, it manifests itself primarily along the Eastern Seaboard, where declining growth rates and the dieback of red spruce and other species of trees, particularly at high elevations, are attributed to atmospheric pollution, of which acid deposition is one of the most widespread components.

A primary human source of the precursors to acid deposition is coal-fired power plants, which provide more than 55 percent of the electricity generated in the United States and account for about one-third of the nitrogen oxides and about two-thirds of the sulfur dioxide produced each year. These atmospheric pollutants are capable of a phenomenon known as "long-distance transport," which simply means that they can travel great distances from their sources on air currents (even as far away as the Arctic and Antarctica) before being deposited on agricultural fields, forests, grasslands, deserts, jungles, and oceans, where they affect plant growth and acidity.[11]

Direct effects of acid deposition include reduced functioning of plant roots. This reduction in root function may be caused by chemical changes in the soil as a result of acid deposition, by reduced translocation of carbohydrates from pollution-damaged shoots, or from excessive nitrogen. In fact, the emission of pollutants has tripled the level of nitrogen some forests are receiving. Change in the level of nitrogen works as follows: People are used to the idea that if one puts nitrogen on a cornfield, the corn grows faster, but then corn happens to be a plant that opens its stomata (pore-like entrances into the leaves) and takes up more carbon to create a balance. Thus, to some point, the more nitrogen corn is given, the more carbon it will acquire, and the better it will grow. Unlike corn, however, most forest species will not grow any faster with additional nitrogen.

Since trees do not take up any more carbon to offset the additional nitrogen, an altered carbon–nitrogen ratio develops in the plant tissue, which means that trees are receiving three times the level of their nitrogen tolerance. This *does not* mean that they are getting three times their optimum level, but rather three times their *tolerance*. Such a shifted carbon–nitrogen ratio translates into, among other things, an alteration of the materials plants produce to resist diseases and insects, which permits pathogenic fungi (those that create rot in living trees) to enter, where previously they had been excluded.[13]

How might this undermine cultural sustainability? People of Canada's First Nations (the various Aboriginal peoples in Canada who are neither Inuit nor Métis) live sustainably within their ancestral lands, which lie within the northern edge of the boreal forest. Their dominant means of livelihood is hunting, fishing, and trapping fur-bearing mammals, as they have done for centuries. Being isolated by miles (kilometers) of unbroken forest and being closely tied to their ancestral lands, they have largely stayed within their borders.

Enter airborne sulfur dioxide from the cities of southern Canada and the northern United States, well over a thousand miles (kilometers) away, and the resulting acid deposition begins in secret to kill the boreal forest. The demise of the forest is at first invisible, but then the people begin to notice that something is wrong. Gradually, the condition of the forest deteriorates to the point that it alters the habitats of the animals on which the people have traditionally relied for food and as money for trade.

With the decline in forest health comes poverty and destabilization of the people's culture within their ancestral home. With nowhere else to go, the culture collapses with the collapse of the forest's biological sustainability, through no fault of the First Nation peoples. In fact, they most likely do not know what happened. All they know is that the forest, whose sustainability they have respected, protected, and relied on for centuries, is suddenly, inexplicably dying, and they are dying with it.

In addition to acid rain, an analysis of tree bark, collected from the tropics to the chilled latitudes, indicates that insecticides and fungicides have spread and are spreading around the world on air currents. Traces of chemicals found in the bark include those related to DDT, lindane, chlordane, aldrin, and more than a dozen other insecticides and fungicides. Some of these chemicals become airborne in hot climates and are carried to cooler areas, where they condense out of the atmosphere and concentrate far from where they were sprayed, often thousands of miles (kilometers) away. Some were even sprayed decades ago but are still affecting the environment through the pollution of soil and water.[14]

Direct and Indirect Pollution of Soil and Water

Unintentional fragility is also imposed on ecosystems through both the direct and indirect pollution of soil and water. Soil, which is like an exchange membrane between the living (plant and animal) and nonliving components of the landscape, is dynamic and ever-changing. Derived from the mechanical and chemical breakdown of rock and organic matter, soil is built up by plants that live and die in it. It is also enriched by animals that feed on plants, void their bodily wastes, and eventually die, decay, and return themselves to the soil as organic matter. Soil, the properties of which vary from place to place within landscapes, is by far the most alive and biophysically diverse part of a terrestrial ecosystem. In addition, soil organisms are the regulators of most processes that translate into soil productivity.

The soil food web is a prime indicator of the productive capacity of a terrestrial ecosystem. But, soil processes can be disrupted by such things as decreasing the biomass of bacteria or fungi and thus their functions, altering the ratio of fungal to bacterial biomass in a way that is inappropriate to the desired system, reducing the number and diversity of protozoa, and reducing the number of nematodes or altering their community structure. Such disruptions can lead to a loss of vegetation or even the loss of human health.[15]

Soil, which is the main terrestrial vessel, receives, collects, and passes to the water all airborne, human-caused pollutants. In addition, such pollutants as chemical fertilizers, fungicides, herbicides, insecticides, rodenticides, and so on are added directly to the soil and through the soil to the water. At times, such pollutants make their way into the air and hence are redistributed more widely over the planet's surface through strong winds, which carry aloft the topsoil following deforestation, desertification, and ecologically unsound farming practices, whereby they ultimately affect water.

Most of our usable water, which is a captive of gravity, comes from snows high on forested mountain slopes. When snow melts, the water percolates through the soil. It is purified when flowing through healthy soil; it is poisoned when flowing through soil stripped of nature's processes and polluted with artificial chemicals. In addition, water bearing tons (metric tons) of toxic effluents flows directly into ditches, streams, rivers, estuaries, and oceans.

Water, the great collector of human-caused pollutants, washes and scrubs the pollutants from the air by rain and snow. It leaches them from the soil, and it carries them in trickle, ditch, stream, and river to be concentrated in the ultimate vessel, the combined oceans of the world.

Humans directly affect air and both directly and indirectly affect soil and water. If, for example, we choose to clean the world's air, we will automatically clean the soil and water to some extent because airborne pollutants will no longer poison them. If we then choose to treat the soil in such a way that we can grow what we desire without the use of artificial chemicals, and if we stop using the soil as a dumping ground for toxic wastes and avoid overintensive agriculture, the soil can once again begin to purify water by filtering it. If we then stop dumping waste effluents into ditches, streams, rivers, estuaries, and oceans, these water bodies can again become clean and biophysically more productive.[16]

With clean air, soil, and water, we can also have clear, safer sunlight with which to power Earth and, with the eventual repair of the ozone shield, a more benign—and perhaps predictable—climate in which to live. In addition, effective population control can tailor human society to fit within the world's biophysical carrying capacity.

A population in balance with its habitat will reduce demands on the global resources. With reduced competition for resources can come the cooperation and coordination that will allow our landscapes to provide the maximum possible biodiversity. Protecting biodiversity translates into the gift of choice, which in turn translates into hope and dignity for *all* generations—present *and* future.

If we did everything outlined here except clean the air, we will still pollute the entire Earth. *Clean air is the absolute bottom line for global social-environmental sustainability—and therefore human survival.* Without clean air, there eventually will be no difference in the way we destroy ourselves because our biosphere is comprised of interrelated, interdependent, interactive components: One affects the whole, and the whole affects the one. But, there are alternatives, based on intrinsic values. The choice is ours. To all generations, beginning with today's children, we bequeath the consequences.

Fifth Element: Understanding and Accepting That We Must Reinvest in Living Systems Even as We Reinvest in Businesses

Most industrialized nations of the world operate with the idea that for an economic endeavor to be healthy, it must be forever expanding. This ultimately results in the overexploitation of most, if not all, "renewable" resources.

The timber industry, a prime example, operates largely in a perpetual expansionistic mode, particularly in areas where considerable native forest remains. As a result, the world's irreplaceable forested resources—such as everything connected with the processes of an indigenous forest—are rapidly shrinking. Perpetual expansion involves liquidating the native forest and ultimately exhausting the soil while concomitantly altering the forest as a water catchment and habitat for its associated species, thus affecting the biodiversity of the forest.

To illustrate, if you look closely at the surface of the floor in an old forest that has not been disturbed by human endeavors, it becomes apparent there is no such thing as a smooth slope. The forest floor is roughened by the scattered stumps, pieces of collapsed snags, and whole fallen trees, their uprooted butts, and the pits and mounds left by their uprooting. Living trees roughen the surface of the forest floor by sending roots outward along slopes, often near the surface. And, tree trunks distort the surface by sloughing bark and by arresting creeping soil at their bases.

Decomposing woody roots of tree stumps also have distinct functions in the forest. Tree roots contribute to the shear strength of the soil, which is a root's ability to hold soil in place without being pulled apart by gravity. Declining shear strength of decomposing woody roots increases mass soil movement (such as a "landslide") after such disturbances as catastrophic fire—or, today, road building and clear-cut logging. Another related function of decomposing tree stumps and roots is the frequent formation of interconnected, surface-to-bedrock channels that rapidly drain water deep into the soil from heavy rains and melting snow. The collapse and plugging of these channels as roots decay can force more water to drain through the soil matrix, reducing soil cohesion, and increasing hydraulic pressure, which in turn can cause mass soil movement.

Nevertheless, the economic concept of waste (considered to be an "economic loss") has spawned the industrial concept of salvage logging, usually in the form of clear-cutting. Clear-cutting in turn is an economic expediency without sound biophysical justification because it mimics nothing in nature.

According to the 1981 *American Heritage Dictionary, salvage* means to save any imperiled property from loss or destruction. Salvage logging, the premise of which views trees—especially old trees, euphemistically referred to as excess "inventory"—only as imperiled economic property, exemplifies the philosophy of traditional forestry, wherein the belief is that any potentially merchantable tree left in the forest to rot is an economic loss, hence a waste. But, in a biological sense, there is no such thing as waste in a forest—or any other ecosystem. *All waste is fuel for other living systems.*

A tree rotting in the forest is a *reinvestment* of nature's biological capital in the long-term maintenance of soil productivity and hence the forest itself. In a business sense, one makes money (economic capital) and then takes a percentage of the earnings and reinvests them in the maintenance of buildings and equipment to continue making a profit by protecting the initial outlay of economic capital over time. In a business, one reinvests after the fact, after the profits have been earned and all incurred expenses have been paid.

It is different, however, with biological capital, which is the capital of all renewable natural resource systems. A forest simply cannot process economic capital. It requires such biological capital as organic material and biological and genetic diversity. In a forest, one must reinvest before the fact by leaving some proportion of the merchantable trees in the forest to rot and become incorporated in the fabric of the living system.

Such biological reinvestment, including large merchantable trees, both alive and dead, is *necessary* to maintain the integrity of soil processes and thus fertility, which in large measure determines the sustainable capacity of the forest to produce the ecological services we humans require. The biophysical integrity of the forest—if honored and protected—can maintain the long-term economic health of the timber industry and therefore those human communities that are largely dependent on it.[17]

In sum, sustainability is additive. We must have a biophysically sustainable ecosystem to have a biologically sustainable yield, a biologically sustainable yield to have an economically sustainable industry, an economically sustainable industry to have a socially sustainable economy, a socially sustainable economy to have culturally sustainable communities, and culturally sustainable communities to have a society blessed with social-environmental sustainability.

When sustainability is put in purely economic terms, the additive economic relationship of the biological yield becomes clear: We must first practice sound *bioeconomics* (the economics of maintaining the biophysical integrity of ecosystems, i.e., a sustainable forest) before we can practice sound *industrioeconomics* (the economics of maintaining a sustainable industry, i.e., forestry) and before we can practice sound *socioeconomics* (the economics of maintaining a sustainable community and society). It all begins with a solid foundation—in this case, the biophysical integrity of a sustainable forest.

In this context, *sustainable* means the production of industrioeconomic outputs while maintaining the viability of forest processes (bioeconomics) in perpetuity. In turn, this necessitates balancing withdrawals of products with bioeconomic reinvestments in the biophysical integrity of the forest, especially the soil. It means maximizing the sustainable integrity of the forest and using all products and enjoying all amenities thereof with prudence, humility, and gratitude.

Sixth Element: Understanding and Accepting That Only a Systemic Worldview Is a Sustainable Worldview

The antecedent to and foundation of the dominant worldview or society's current operating paradigm of how the world works was fashioned by such rationalist thinkers as Francis Bacon (1561–1626, English philosopher and essayist); Galileo Galilei (1564–1642, Italian scientist and philosopher); René Descartes (1596–1650, French philosopher and mathematician); John Locke (1632–1704, English philosopher); Isaac Newton (1642–1727, English mathematician, scientist, and philosopher); and Adam Smith (1723–1790, Scottish political economist and philosopher). Consider the collective paradigm of these renowned men: Nature's sole value is in service to the material desires of humanity (Bacon). But, nature must be tortured before its secrets will be revealed for human use (Bacon).[18] Once wrested from nature, only those secrets that are measurable and quantifiable are real or relevant and can be studied (Galilei). Because real things are both measurable and quantifiable, they must operate through predictable linear mechanical principles, like an enormous machine (Newton). And, like a machine, real things can be understood by disassembling the things themselves into smaller and smaller, more manageable pieces, which can then be rearranged in an order deemed logical to the human mind (Descartes).[19]

With reductionistic, mechanical logic, major segments of Western society confer on themselves the unlimited rights of individual private property (a distortion of Locke's original intent) for which people must compete with one another in pursuit of their own self-interests (Smith). (*Reductionistic* is used in the sense of taking apart and isolating the components of a living system, rearranging them according to human logic, keeping only those deemed of value in human terms, reassembling the retained pieces, and expecting the system to function as before.[20] This is much like dissecting a living being, keeping the parts for which one sees a value, arranging the parts in a humanly conceived order, sewing them together, and expecting the organism to live and behave like it did. *Mechanical* refers to the common human notion that the world is assembled like a machine, acts like a machine, and thus can be treated like a machine with interchangeable parts, like a watch.)

Such self-interest is to be free from any government interference because the "invisible hand" of moral guidance will temper self-interest in the pursuit of material wealth—for the betterment of society (Smith). While Smith's invisible hand may have had spiritual connotations, they are ignored in the current pursuit of self-interests in the form of material wealth. Further, his notion of a higher moral guiding human action was already overshadowed by the accepted reductionistic mechanical posits of Bacon, Galileo, Descartes, and Newton.[21]

The legacy of the reductionistic, mechanical foundation of an expansionist-economic worldview has led to many of our most serious human-created problems and has also fed people's sense of isolation from one another and from nature. This dualism has led us to treat nature as a commodity from which we are independent and separate. By intellectually isolating ourselves from nature, we have justified our trying to control the uncontrollable.

Our analytical perspective involves a four-part process: (1) dissect the system into its component parts, (2) study each part in isolation, (3) glean knowledge of the whole by studying its parts, and (4) rearrange the parts in such a way that they satisfy a human sense of logic aimed at ever-expanding growth and capital gain.

The implicit assumption is that systems are aggregates of interchangeable parts that function in a linear fashion. Thus, by optimizing each part, we optimize the whole. We therefore continually fragment our problems into smaller, more "manageable" pieces, while our challenges are increasingly systemic. (*Linear,* in the sense it is used here, means having only one dimension, that of an ever-extending straight line with no means of return, such as an economy that is ever expanding.)

Consider, once again, a forest. A forest is dissected into its components as best we are able, such as trees, shrubs, herbs, birds, mammals, water, soil, and so on. Each component is then studied in relative isolation of the others, often in terms of its perceived economic potential. With considerable knowledge of those parts we favor for their potential economic value and almost no knowledge of those parts for which we find no immediate use, we cobble together our sense of the whole as a functioning system. Finally, we select and arrange the pieces for which we perceive an economic value in the order that we think will accrue the greatest financial benefits, in the shortest possible time, at the least possible cost, into an unlimited future. And behold: We have converted a biologically sustainable, complex forest into a biologically nonsustainable, economically simplistic tree farm.

Today, fragmentation based on linear, reductionistic, mechanical thinking, which looks at parts and ignores the whole, fails to see society as a living system. This kind of fragmentation led quantum physicist David Bohm to say: "Starting with the agricultural revolution, and continuing through the industrial revolution, increasing fragmentation in the social order has produced a progressive fragmentation in our thought."[22] Fragmentation—specialization, special interest groups, and political lobbyists—is the very underpinning of today's professionalism, yet it is making our society increasingly ungovernable. The triumph of such reductionist thinking has given rise to a whole set of economic conditions under which we try to operate in isolation from the system itself.[23]

Although sustainability is only possible with the acceptance of a unified worldview, management, which is a reductionistic, mechanical concept, is the antecedent of an expansionist-economic worldview. Plans to manage Earth are therefore founded on the belief that ignorance is a problem that can be cured with science and technology, both of which are linear, reductionistic, and mechanical. And, the purpose of so-called land management is to benefit an ever-expanding economy.

A worldview has incredible power in directing the way people think. In addition, assumptions embodied in a worldview become "second nature" and therefore part of our subconscious. We perceive them to be absolute truths when in reality they are illusions accepted as truths. We must therefore examine carefully our present, obsolete, expansionist-economic worldview and begin a conscious transition to a worldview with a far greater sense of social-environmental sustainability (a unified, systemic worldview) if our society is to survive the twenty-first century in any semblance of its current order.

The Transition

For a new worldview to emerge, its creators must be willing to take the time to identify, clarify, and evaluate the underlying assumptions of the existing dominant worldview before the transition to a new worldview is possible. Failure to do so ensures that the power of habit and the tyranny of the familiar values of the old model will co-opt those of the fledgling view, according to Professor Duncan Taylor.[24]

Taylor goes on to contrast the current expansionist-economic worldview with an emerging ecological worldview. To wit, just as the expansionist-economic view excludes, to the greatest extent possible, the ecological lessons of history and current scientific knowledge, so the ecological view excludes, to the greatest extent possible, economics in the social sense.

Both views have thus become what they are against—being exclusionary. This is ironic because ecology and economics have the same root: *eco* from the Greek *oikos*, a house. Ecology is the knowledge of the house, economics is the management of the house, and it is the same house. Both views thus ignore the integrity, vitality, and sustainability of the whole.[25]

The way out is a worldview that brings the material and the spiritual into harmony, heals the human/nature schism, simultaneously creating a mutualistic symbiosis between ecology and economics.

Such a view must recognize the value of relationship and accept that the only way anything can exist is encompassed in its interdependent and interactive relationship to everything else, which means that an isolated fragment or an independent variable exists only on paper. As such, every relationship is dynamic, constantly adjusting itself to fit precisely into every other relationship, all of which are consequently adjusting themselves to fit precisely into every other relationship, and so on ad infinitum.

Can so fluid a notion as ever-adjusting relationships be made to work within a rigid, mechanistically oriented, reductionistic social construct powered by the notion of ever-increasing economic expansion? The answer is no because, through self-reinforcing, behavioral feedback loops, our present social paradigm protects the status quo by condemning change as a condition to be avoided at almost any cost.

While no one is safe from the many environmental problems engendered by our current worldview, which threaten our health and biophysical integrity of the planet, we are not all endangered equally. In addition, the perceived security we have long sought through ever-increasing consumption, militaristic technology, and domination over nature has actually threatened our long-term social survival. This threat prompted Czech President Vaclav Havel to observe: "Without a global revolution in human consciousness a more humane society will not be possible."

We must, through a global revolution in human consciousness, change our worldview (from one based strictly on a linear commodity-oriented thought process) to a systems approach in which the indicators of both ecological and economic sustainability are rooted in the quality of the relationships between and among the parts. Such a view does not mean that the linear commodity-oriented thought process is summarily discarded. Rather, it means that the linear commodity-oriented thought process is moderated and ultimately constrained by those biophysical limitations that control the sustainability of social-environmental harmony.

If we have the moral courage and the political will to adopt and implement the concept of sustainability as a component of social evolution in which change is accepted as a process to be embraced, then the notion of ever-adjusting relationships becomes the creative energy that guides a vibrant, adaptable, ever-renewing society through the present toward the future. And, because sustainability honors the integrity of both society and its environment, the outcome is a unified worldview in which the *function* of a system defines the system. That is, the function defines the composition, which in turn defines the structure, and it is after all the function that either sustains social necessities or eliminates them.

The unified, systemic worldview is predicated on the notion of holism, in which reality consists of organic and unified wholes that are greater than the simple sum of their parts. The following are basic assumptions on which the unified, systemic worldview is founded: (1) Everything exists in relationship to everything else; (2) every relationship is dynamic, constantly adjusting itself to fit precisely into all other relationships, which consequently are adjusting themselves ad infinitum; (3) all relationships, including nonmonetary ones, have value; (4) everything, including humans and nonhumans, is interconnected, interdependent, and interactive; (5) all relationships are systems supporting systems supporting systems ad infinitum; (6) the whole is functionally greater than the sum of its parts; (7) processes have primacy over components; (8) a system is defined by how it functions, not by the shape, number, or arrangement of its component; (9) the integrity of the environment and its ecological processes has primacy over human desires when such desires would destroy the system's integrity for future generations; (10) nature determines the limitations of human endeavors; and (11) the relevancy of knowledge depends on its context.

In the unified, systemic worldview, the behavior of a system depends on how individual parts *interact* as functional components of the whole, as opposed to what each isolated part is doing, because the whole is understood through the relation/interaction of its parts. To understand a system, therefore, we need to understand how it fits into the larger system of which it is a part. This gives us the view mentioned of systems supporting systems supporting systems ad infinitum. We thus move from the primacy of the parts to the primacy of the whole, from insistence on absolute knowledge as truth to the relatively coherent interpretations of constantly changing knowledge, from isolated self to self in community, and from attempting to solve old problems to creating new concepts.

In a unified, systemic worldview, the disenfranchised as well as future generations have constitutional rights that must be accounted for and honored in present decisions and actions. In addition, individual people—as well as their relationships among one another, nature, and their communities—both have value and are valued, as are all living beings. It is this combined sense of feeling valued as a person and feeling the personal value of contributing positively to one's community that makes sustainability a workable proposition for the present into the future.

Seventh Element: Accepting Our Ignorance and Trusting Our Intuition While Doubting Our Knowledge

There is much to learn about Earth and from Earth, and there are many good reasons to believe that its complexities are permanently beyond our comprehension. Thus, the salient point is not our knowledge but our ignorance because only our ignorance can be proven. The credibility of our knowledge rests, albeit tenuously, before the jury of tomorrow and the day after that and the day after that.

For the present, however, it is necessary to keep the search for truth on its own credible track. This requires that we recognize our tendency not only to form a single hypothesis and become so attached to it that any criticism of or challenge to our methods raises our ego defenses but also our tendency to "massage" the theory to fit the data and of the data to fit the theory.

Moreover, while scientists may attempt to detach themselves from nature and to become *objective,* they are never completely successful since they are part of nature and must participate with nature to study nature. Even history, which we tend to think of as fact, is viewed individually through interpretation of that which we perceive: "My view of history," says historian Arnold Toynbee, "is itself a tiny piece of history; and mainly other people's history and not my own."[26] Nevertheless, the role of science is to help humanity understand the biophysical principles governing the universe and how we are related to them as an inseparable, functioning part of that universe.

The irony of scientific research is that nothing can be proven; it can only be disproved. Therefore, we can never "know" anything in terms of knowledge. We can know only in terms of intuition, which is the knowing beyond knowledge and not admissible as evidence in modern science. Whatever truth is, it can only be intuited and approached, never made concrete.

Knowledge, therefore, which is external to a person, is not "knowable," and intuition, which is internal to a person, is not knowledge and therefore is not subject to disproof. Intuition is inner sight—individualized, inner knowing that is beyond knowledge—for which proof is unnecessary and explanation impossible.

Intuition, a mode of knowing widely accepted since ancient times, has been clouded with ambiguity and controversy since the advent of the reductionistic-mechanical mindset, which swept Western society more than a century ago. For some, intuition is merely a meaningless by-product of unconscious processes, but for others intuition, an instantaneous, direct grasping of reality, is the source of our deepest truths, that sense of unquestionable knowing, of which even John Stuart Mill, a pillar of the empirical method, said: "The truths known by intuition are the original premises from which all others are inferred."

Intuition is therefore more than a guide to truth, according to sociologist Pitirim A. Sorokin. It seems to be the foundation of our conception of the beautiful and the good and thus the basis for our aesthetic and moral judgments.[27]

It is also well known that there are, among the founders of quantum physics, a number of self-avowed mystics and intuitionists, such as Niels Bohr, Albert Einstein, Sir Arthur Eddington, and Eugene Wigner. To them, such principles as time, space, and the conservation of forces are deep intuitions, if not acts of faith.

Knowledge, on the other hand, is a collective outer experience of humanity and society's subjective judgments about things that is shareable, but only relative, and disprovable. Thus, the actual objects of our inquiries, the formulations of our questions and definitions, and the mythic structures of our scientific theories and facts are social constructs. Every aspect of our scientific theories, facts, and practices—including scientific method itself—are but expressions of contemporary socio-/political/economic interests, cultural themes and metaphors, personal biases, and personal/professional negotiations for the power to control, albeit momentarily and minutely, the scientific knowledge of the world.

Facts that scientists construe to be statistically true statements about nature are presented as concrete, often irrefutable, outcomes of a social process of developing statements about the world to distinguish order from perceived chaos. Thus, instead of scientific consensus being achieved when the *facts* reach the state of "speaking for themselves," scientists come to a consensus when the political, professional, and economic costs of refuting them make further negotiation untenable.[28]

There is, however, no single reality but rather a multiplicity of realities, the representation of which depends on one's position in the process of negotiating an acceptable social view of reality. Thus, regardless of the question, the reproducibility of the experimental design and methods does not mean that the results represent anything necessarily true about nature. The reproducibility of the experimental design shows merely that a particular negotiation of reality is reproducible under a certain set of conditions.

The problem is that we confuse the limits of scientific method with the nature of ordinary experience. While scientific method tends to be reductionistic and mechanical, life experiences are synthetic holograms made of complex threads of past experiences, present perceptions, and dreams of future opportunities and fears of future circumstances. Life is thus a moment-by-moment kaleidoscope of a perceptual present in eternal flux within and between internal and external realities.

If, therefore, we are going to ask intelligent questions about the future of Earth and our place in the scheme of things, we must understand and accept that most of the questions we ask deal with cultural values, which cannot be answered through scientific investigation. Nevertheless, scientific investigation can help elucidate the outcome of decisions based on these values and must be so employed. In addition, we might be wise to consider the gift of Zen and approach life with a beginner's mind—a mind simply open to the wonders and mysteries of the universe.

A beginner sees only what the answers might be and knows not what they should be. If, on the other hand, a person perceives himself or herself as an expert, that person may think he or she knows what the answers should be and can no longer see what they might be. The beginner is free to explore and to discover a multiplicity of realities, while the expert typically remains loyal to pet realities, which often turn into notions to be protected at any cost. The beginner's mind thus frequently understands the question better than the expert does.

If we are going to ask intelligent questions about sustainable land use planning, we must be open to multiple hypotheses and explanations, and we must be willing to accept a challenge to our ideas in the spirit of learning rather than as an invitation to combat. The greatest triumphs in science are not, after all, triumphs of facts but rather triumphs of new ways of seeing, of thinking, of perceiving, and of asking questions. To achieve sustainability, therefore, we must learn to accept our ignorance and trust our intuition while doubting our knowledge.

Eighth Element: Specifying What Is to Be Sustained

The meaning of *sustainability* is clear in that it is consistently used, either explicitly or implicitly, to mean continuity through time. Whatever ambiguities exist are not with the concept of sustainability itself, but rather with the implications of sustainability in a given context. For example, what is the scale of time within which something must exist to be considered sustainable or sustained? This question does not impinge on the meaning of sustainability itself as a mark of continuity; it is only a matter of duration or scale.

To sustain anything, one must specify what is to be sustained and what is not. This is an important concept because when one elects to sustain one thing, one simultaneously elects not to sustain something else.

For example, the language of the Multiple Use Sustained Yield Act of 1960, although of good intent, is based on an economic assumption, which is totally at odds with ecological reality. The assumption is that biological processes in a forest ecosystem remain constant while we humans strive to maximize whatever product or amenity of the forest seems desirable.

Sustained yield is thus an economic concept in which the volume of wood fiber to be cut annually is predetermined by the economic targets of industry. These targets have nothing to do with the ecological capability of the forest to produce that volume on a sustainable basis without impairing its ability to function ecologically over time. *Sustained yield* signifies that the Congress of the United States has chosen to sustain the yield—rather than the biological processes of the forest that produces the yield.

If, however, the real intent of Congress had been to maintain forest productivity, consider the following question, which represents a fundamental shift in perception: How would we be thinking about and acting toward our forests, especially our national forests, if the Multiple Use Sustained Yield Act had been conceived of and written as the Multiple Use *Sustainable* Yield Act or even as the Multiple Use *Sustainable Forest* Act?

In the last two cases, the sustainability of the forest processes and long-term productivity of the soil would be the object of maintenance—rather than the yield of wood fiber. This choice would have protected the system's biophysical sustainability and thereby ensured sustainable commodity production rather than ensuring short-term economic commodity output at a predetermined level at the cost of the system's long-term productive capacity.

While sustainability does not exclude the extractive use of resources, it does demand a balanced approach to their long-term maintenance, extraction, and use. This means that the economic divestment of resources from any ecosystem must be balanced by the biological reinvestment of resources in that same system, regardless of the economic impact on the profit margin—something that is not now happening. The concept of balance is important because we live between two spheres: the atmosphere (air) and the hydrolithosphere (the continents and the water surrounding them). If we destroy either one, we will be the authors of our own extinction, which posits a question: At what scale is sustainable development possible?

Although we, as a society, do not have many answers, we can find guiding principles for action in the questions we ask. For example, thinking about the scale at which sustainable development is possible calls to mind a catchphrase: Think globally; act locally. But, when considering the notion of global community from an individual's standpoint, helping to heal the world seems like a hopeless task, even an unintelligible abstraction. Yet, as people come to understand their effect on Earth, an increasing number of them want to do something to change the worn-out expansionist-economic paradigm. But, if global and national strategies are abstractions too far removed from the average person's realm of experience, at what level can the average person act and be effective?

Sustainable development must be implemented where people are invested in a sense of place and are able to learn, feel, and be empowered to act—the local level. Sustainable development must be integrated into policies and decisions in local communities, where people have the power to effect change and make decisions based on a systemic worldview, a "first-world" view: Act locally and affect the whole world. Then, one begins healing the environment in the present for the future.

Here, as he so often does, author Wendell Berry cuts to the core: "That will-o'-the-wisp of the large-scale solution to the large-scale problem, so dear to governments and universities and corporations, serves mostly to distract people from the small, private problems that they may in fact have the power to solve. The problems, if we describe them accurately, are all private and small. Or they are so initially."[29] It is thus imperative that we address the fundamental causes of our problems at their roots—our thinking and behavior at the level of the local community—or we will always be dealing with symptoms and Band-Aid solutions that compound the problem by denying the cure.

Through the concept and practice of local community, people can empower themselves and support one another through decisions that promote a sustainable world as well as increase their quality of life. Within a local community, people can act as the force driving change in their political system even as they alter their lifestyles. Through their actions, people partake in guiding destiny, despite the fact that a local community is part of a larger, more impersonal governmental system.

Local community governments, as opposed to county, state, or national governments, have both a greater degree of understanding and the interaction necessary to amend local problems, such as land use, waste reduction, political representation, and education. They are therefore better able to implement and adjust to aspects of sustainability in the social-environmental arena.

People both define their local communities and are defined by them in that communities play a primary role in maintaining cultural values within and among generations. The collective of individual values determines familial values; the collective of familial values determines community values and determines appropriate behavior, poverty, and success. As we grow up and are taught at home, educated in schools, and participate in community, socialization occurs, norms are set, and societal control takes place.

People make most of their decisions, do most of their consuming and waste production, and develop many personal relationships within their local community. It is not surprising therefore that lifestyle becomes a political issue.

Citizens at the local level can also begin drawing connections between personal consumption and its effects on local, regional, national, and global economic well-being and environmental integrity. Although political pressure must be exerted continuously on national governments, both at home and abroad, lifestyles in each and every local community have direct and immediate effects on the biosphere.[30]

Local communities, through their collective effects on world society, act as catalysts for change in society at the local, regional, state, and national levels and finally in the world itself. Through the behavior of their individuals, local communities contribute greatly to environmental integrity and the global climate. And, because we, as individuals, collectively comprise local communities, which in the collective comprise society at the regional, state, and national levels, we can begin to repair our global environment simply by changing our individual behaviors. Local communities, as the force that drives change for better or worse, are thus the appropriate scale for dealing with sustainable development.

Ninth Element: Understanding and Accepting That Sustainability Is a Continual Process, Not a Fixed End Point

Sustainable development is based on the recognition that sustainability is a continual process centered in conscious, purposeful, cultural evolution (both personal and social), as opposed to a condition that once achieved becomes static and thus can be ignored. The outcome of such transformative evolution must be a self-reinforcing feedback loop of ever-growing, social-environmental harmony as people attain psychological maturity.

Because of the dynamic nature of the evolving ecosystems we attempt to manage, such as a forest, grassland, or ocean, we can only *attempt* to "manage" (= control) an ecosystem for its evolution—not for a sustained yield of products as an economic absolute. Therefore, the only sustainability we may be able to achieve is that which ensures an ecosystem's ability to adapt to changing conditions, such as warming of the global climate.

In addition to nature's disturbances through which ecosystems evolve, they are also influenced by human-introduced disturbances. In the early years of the nineteenth century, for example, a view emerged that strong interdependencies among the climate, the plants and animals, and the soil led to the long-term stability of terrestrial ecosystems in the context of landscapes. But, this notion was considered only in the context of a stable climate on a spatial scale. Today, a revised concept is emerging, which might be stated as follows: The spatial patterns of ecosystems, including those of forests or grasslands, observed on landscapes result from complex interactions among physical, biological, and social forces over time.

Because most landscapes have been influenced by the cultural patterns of human use, the resulting landscape is an ever-changing mosaic of nonmanipulated and manipulated patches of forestland, shrubland, grassland, and other habitats, which vary in size, shape, and arrangement. This spatial patterning is a unique phenomenon that emerges at the level of the landscape that is constantly changing throughout time.

With respect to human-introduced disturbances, such as fragmentation of habitat, biogeographical studies have shown that the connectivity of the areas within a landscape is important to the persistence of ecological processes as well as plants and animals in viable numbers in their respective habitats. In this sense, the landscape can be considered as a dynamic interplay of connected patches of habitat, such as forested riparian areas that act as corridors or routes of travel between patches of upland forest or other suitable habitat.[31]

To allow and help ecosystems, such as forests, migrate in the face of global climate change, we must emphasize diversity in all of its aspects, which in turn will do much toward ensuring the resilience and sustainability of those systems. We also must pay close attention to the patterns we create on the landscape because the overall connectivity of those patterns will either allow or inhibit the ability of species, biological communities, or even whole ecosystems to migrate with changing climatic conditions.

The ability of a species, biological community, or even a whole ecosystem to migrate will determine how much of the genetic variability—adaptability—is lost from the gene pool with the extinction of local populations of plants and animals. Such extinctions inevitably accompany the fragmentation of ecosystems within landscapes and across biogeographical regions. Part of the reason these extinctions have thus far been inevitable is that, while people think saving species is important, they have little or no understanding of the cause and effect wrought through the aggregate of human activities on the health of habitats and the environment as a whole.

What decisions we make are up to us, but one thing is clear: While the current trend toward homogenizing ecosystems may make sense with respect to maximizing short-term profits, it bodes ill for the long-term ability of both ecosystems and landscapes to adapt to changing environmental conditions on a sustainable basis. We suggest, therefore, that while people must strive in the present to manipulate ecosystems on a biologically sustainable basis, they must simultaneously recognize that the concept of sustainability may prove to be relatively shorter lived than anticipated. We say this because the climate is changing—despite the fact that it has long been taken for granted as a constant by traditional economists and thus omitted from economic models.

As evidence of global warming grows, it becomes increasingly apparent that climate must now be accepted and accounted for as a variable of great uncertainty. In the realm of land use planning, we must now look beyond the sustainability of ecosystems as isolated entities in time and space to their long-term sustainability as contextual components of landscapes whose adaptability to changing environmental conditions must be protected.

The long-term sustainability of ecosystems may well depend on the connectivity of the patterns we create on the landscape and the adaptability of those patterns to the forced migration of ecosystems (or components thereof) over time as demanded by changing environmental conditions. This means that we must rise to the challenge of manipulating ecosystems first and foremost for the sustainability of their internal processes; second as adaptable patterns on rapidly changing landscapes, which may shortly be subjected to unprecedented global warming; and third for the goods and services they provide to humanity.

With these priorities in mind, our responsibility in making current decisions about the sustainability of the adaptive processes entrained with the patterns we create across landscapes is to be mindful of the consequences our decisions will have on future generations and their ability to respond to the conditions that our decisions will create. This is particularly important because economic and ecological systems are perceived to operate on different scales of time, which means that the long-term detrimental effects of decisions made in favor of short-term profits are ignored. As a result, the long-term, nonmarket value of nature's services is grossly underestimated.[32]

Where we, as a species, will ultimately end up, with an environment compatible with our existence or hostile to our existence, depends on the level of consciousness we bring to the questions we ask. And, before we can obtain fundamentally new answers, we must be willing to risk asking fundamentally new questions.

We must therefore look long and hard at where we are headed with respect to the patterns we create on the landscape as we move from the most natural end of the landscape continuum toward the most cultural end. This is important because the old questions and the old answers have both gotten us to where we are today in the fragmentation of ecosystems and are guiding us to where we will be tomorrow in accelerating such fragmentation toward the ultimate collapse of ecosystems.

We can, however, manipulate our landscapes so that we consciously protect, maintain, and create as many options as possible both for the generations to come and for ourselves. To do this, we must understand that the quality of the question is all important.

What is out there? What do we want out there? What are our values? Is what we want ecologically compatible with the landscape, as it has evolved over the millennia? What question must we ask to ensure that we maintain as many ecologically viable options as possible, both now and for the future? How must we behave to live compatibly with the land? These types of questions must be asked continually and simultaneously with our decisions about what and how to manipulate an ecosystem, and each question must be answered to the absolute best of our ability, even as we redesign the landscape in an attempt to suit our cultural desires and necessities.

We must also understand that conflicts over values, either ecological or human, are not battles over numbers but rather battles over different worldviews and their respective long-term ecological sustainability. We cannot, after all, legislate feelings or values, only behavior. In the end, it is the sum of the consciousness of our daily choices that will make the difference, choices that ultimately are reflected in the questions we ask.

Thus, by its very nature, the sustainability of a dynamic system is a moving target. Although it may be more clearly sighted with increasing spiritual development and greater scientific/social knowledge, sustainability will never be a finite end point in an ever-evolving world.

Tenth Element: Understanding, Accepting, and Being Accountable for Intergenerational Equity

Intergenerational equity means the responsibility of the current generation not only to its own members but also to all generations to come. The concept of intergenerational equity or environmental justice, from the human point of view, asserts that we owe something to every other person sharing the planet with us, both those present and those yet unborn. But, what exactly, you may ask, do we have to give?

There are certain things that all humans can give whatever their circumstances: love, trust, respect, and the benefit of their experiences. These are the essence of human values that we either extend to one another or withhold. These are the values embodied in each and every option we pass forward or withhold in each and every option we foreclose. When everything is said and done, all we have to give the children of today, tomorrow, and beyond is options and some things of value from which to choose.[33]

Eleventh Element: Understanding, Accepting, and Being Accountable for Biophysical Limitations to Land "Ownership" and the Rights of "Private Property"

Private ownership of land is a recent concept. For example, the pygmies of central Africa, the most ancient of all forest dwellers, hold no enforceable claims to the forests they have inhabited for at least 40,000 years.[34] Indigenous peoples on every continent find the notion of private ownership of land to be both ludicrous and impossible.

How can an individual human being own something that he or she has not created and therefore cannot control? How can an individual own something that has been around for millennia before he or she was born and will continue for millennia after he or she has passed on? How can an individual own something that is so obviously part of the global commons in both time and space that it belongs to every living creature in its turn and so to no one individual in particular at any given time? How is it that we in the United States *cannot* legally steal from our adult neighbors in the present (= in space), but we *can* legally steal from our child neighbors in the future (= in time) through nonsustainable overexploitation of resources? Society is divided on this issue. Society must therefore decide. Either this peculiarly human notion of land ownership and the rights of private property will continue or people must accept an alternative, such as that of custodial trustee of the piece of Earth one inhabits or otherwise has deed and title to as a living trust for the beneficiaries of the future. It is a question of self-centeredness versus other-centeredness, which means that the morality of the idea of land ownership and unlimited rights of private property must be opened to rigorous debate. And, from our perspective, the question of land ownership and the rights of private property is, at its very root, a moral one.

In such a debate, the questions must be: Does the holder of a deed to land of any kind "own" the land in an absolute sense, or is he or she only a custodial trustee? Does such a holder of a deed have the moral right to degrade the productive capacity of the land before passing it on to the next person who must use it? Put a little differently: Does any person have the moral right to steal options from the future for immediate personal gain by irreparably degrading the productive capacity of the land? If not, how can one be granted the legal right to do so?

These questions are addressed in a proposed "common property" amendment to the U.S. Constitution, proposed by Ojibwe leader Walt Bressette (1947–1999) in the early 1990s. This amendment, also known as the Seventh Generation Amendment, does not abrogate the current constitutional protection of property rights (Fifth Amendment) but rather adds to the definition of property. Common sense tells us that common property exists, is being undermined, and needs protection, just as private property has protection. The proposed amendment thus reads: "The right of citizens of the United States to use and enjoy air, water, wildlife, and other renewable resources determined by the Congress to be common property shall not be impaired, nor shall such use impair their availability for the use of future generations."[35]

Should the status quo prevail, however, it will do so because of what Professor David Orr calls "conservatives against conservation." The discussion that follows is taken from his insightful article.[36]

"The philosophy of conservatism has swept the political field virtually everywhere," says Orr, and virtually everywhere conservatives have forgotten what conservatism really means. "What is conservative," asks Orr, "about squandering for all time our biological heritage under the pretext of protecting temporary property rights?" Present-day conservatives scorn efforts by the public to protect such things as endangered habitats (like old-growth forests); endangered species (like spotted owls, coho salmon, and red wolves); clean air; and clean water. Almost any restriction placed on the rights of an individual to use land as private property is being viewed increasingly as an unlawful "taking," even when such use would irreparably damage the land and its surrounding environment. How, one might ask, is it any more of a *lawful* taking when one degrades land in the present that must be used in an impoverished condition by someone in the next generation and beyond?

Even John Locke, from whom we have derived much of our land use law and philosophy, said that, "Nothing was made by God for Man to spoil or destroy." "The point," says Orr, "is that John Locke did not regard property rights as absolute even in a world with a total population of less than one billion, and neither should we in a world of 5.7 billion."

"What," asks Orr, "is conservative about conservatives' denial of the mounting scientific evidence of impending climatic change?" Climate change will have rapid, self-reinforcing feedback loops that could change the nature of Earth's hospitality to human life for all time. What right do we have to run such a risk when the consequences belong to young generations of today and all generations of the future, and they have no voice in either the decisions or their outcomes?

What, asks Orr, is conservative about perpetual economic expansion when it not only has changed Earth more radically than any other force in modern times but also is rapidly destroying communities, traditions, cultural diversity, and whole ecosystems throughout the world? What is conservative about passing forward a despoiled legacy to the future?

Edmund Burke put the capstone on land ownership and the unlimited rights of private property as a sustainable proposition when he wrote:

> Men are qualified for civil liberty in exact proportion to their disposition to put moral chains upon their own appetites. ... Society cannot exist unless a controlling power upon will and appetite be placed somewhere, and the less of it there is within, the more there must be without. It is ordained in the eternal constitution of things that men of intemperate minds cannot be free. Their passions forge their fetters.[37]

Human Relationships

Everything must exist in relationship to something else; conversely, nothing can exist out of relationship to something else. Relationship, the quintessential thread running through time and space, is the essence of sustainability. It is also the essence of humanity because we are creatures who must share to find value, regardless of how it is defined.

For example, what is your favorite place on Earth? How would you feel if you owned it outright? How would you feel if you owned it outright but were the last person on Earth and had no one with whom to share the experience? How would you feel if you owned your favorite place outright and had one other person with whom to share it? Two other people? Three other people?

Mahatma Gandhi, in a conversation with his grandson, Arun, a few weeks before his assassination, handed him a talisman on which he had engraved "Seven Blunders." From these seven blunders, said Gandhi, grows the violence that plagues the world:

1. Wealth without work
2. Pleasure without conscience
3. Knowledge without character
4. Commerce without morality
5. Science [and technology] without humanity
6. Worship without sacrifice
7. Politics without principles

Gandhi called these imbalances "passive violence" and said that, "We could work till doomsday to achieve peace and would get nowhere as long as we ignored passive violence in our world."[38] To his grandfather's list of seven blunders, Arun added an eighth: rights without responsibility.[39]

That, says Donella Meadows, was in 1947, almost 50 years ago, and today every blunder remains institutionalized, built into our corporations, our government, and our very culture. In fact, she says, we actively practice them and even take pride in some. She then goes on to add seven more blunders:

1. Justice without mercy
2. Order without freedom
3. Talking without listening
4. Stability without change
5. Private interests without public interests
6. Liberty without equality
7. Or, in every case, vice versa

If, as Meadows points out, you suggest putting morality into commerce, you are accused of being against commerce. Question wealth without work and you are a jealous hater of rich people or out to undermine capitalism, says Meadows.[40] Whenever we make our private or public discussions about relationships an either/or proposition, we lock ourselves into an unwinnable position of defending a simplistic point of view that has little or nothing to do with the larger reality beyond that of our own perception.

Relationships are the strands in the web of life, and the web is inescapable. For this reason, the following human relationships are central to sustainability: (1) intrapersonal, (2) interpersonal, (3) between people and the environment, and (4) between people in the present and those of the future.

Intrapersonal

An *intrapersonal* relationship is the relationship that exists within a person. It is the individual's sense of his or her own spirituality, self-worth, personal growth, and so on. In short, it is what makes that person conscious of and accountable for his or her own behavior and its consequences. The more spiritually conscious one is, the more other-centered one is, the more self-controlled one's behavior is, and the greater one's willingness to be personally accountable for the outcome of one's behavior with respect to the welfare of fellow citizens, present and future, and Earth as a whole.

Interpersonal

Interpersonal relationships are those that take place between and among people. For example, if someone in a store is rushing blindly to get somewhere and shoves you out of the way, you have a choice in how you respond to being shoved. You can become angry and impatient and say something nasty, or you can be patient, kind, and understanding. Your thoughts and actions are the seeds you sow each time you make a choice.

We always have a choice, except in choosing. *We must choose.* If we do not like the outcome of our choice, we can always choose to choose again. We are not, therefore, victims of our circumstances but rather consequential products of our choices. And, the more we are able to choose love and peace over fear and violence, the more we gain in wisdom and the more we live in social-environmental harmony. This is true because what we choose to think about determines how we choose to act, and our thoughts and actions set up self-reinforcing feedback loops or self-fulfilling prophecies that become our individual and collective realities.

Another tendency of human beings faced with a perceived threat to their sense of material survival is to defend a point of view. There are, however, as many points of view as there are people, and everyone is indeed right from his or her perspective. Therefore, no resolution is possible when each person is committed only to winning agreement with his or her position.

The alternative is to recognize that "right" versus "wrong" is a judgment about human values and is not a winnable argument. It is best, therefore, to define the principle involved in the discussion as the fundamental issue and focus on it. An issue, usually perceived as a crisis, becomes a question to be answered, and in struggling toward the answer, both positive and negative options become not only apparent but also a choice that must be made.

Fear, a projection into the future of unwanted possibilities, breeds weakness, a state in which there is little time or energy left to develop other areas of life. Out of the weakness of fear, humans too easily and too often turn to war in an effort to assert what little power they think remains to them.

The instantaneousness of today's news does not give us time to assimilate the stories within the context of global proportions. News came more slowly in the past and could more readily be kept in context relative to the time and area covered by the information. Today, however, newsworthy disasters all seem to happen instantly in our homes via television and radio and can become so overwhelming that we become numbed emotionally even as they augment our fear of our own unknowable future. In addition, insurance companies continually foster a disaster mentality.

For sustainability to be possible, a certain fearlessness and other-centered or community-centered mindset is critical. In a truly sustainable community, true democracy is at work. Although egos and self-centeredness may dominate for a time, a better understanding of what democracy really means may alter this pattern. As Terry Tempest Williams puts it, "The human heart is the first home of democracy. It is where we embrace our questions. Can we be equitable? Can we be generous? Can we listen with our whole beings, not just our minds, and offer our attention rather than our opinions? And do we have enough resolve in our hearts to act courageously, relentlessly, without giving up—ever—trusting our fellow citizens to join with us in our determined pursuit of a living democracy?"[41] Teamwork demands the utmost personal discipline of a true democracy, which is the common denominator for lasting success in any social endeavor.

Between People and the Environment

Sustainability means that development programs must, to the extent possible, integrate the local people's requirements, desires, motivations, and identity in relation to the surrounding landscape. It also means that local people—those responsible for development initiatives and their effect on the immediate environment and the surrounding landscape—must participate equally and fully in all debates and discussions, from the local level to the national.[42] Here, a basic principle is that programs must be founded on local requirements and cultural values in balance with those of the broader outside world.

Some time ago, I (Chris) attended a meeting on the development of rural communities at which economic diversification was the sole focus of discussion. It soon became apparent that the group was unaware of the importance of landscape to the identity of a community. For example, a logging community is set within a context of forest, a ranching community is set within the context of lands for grazing, and a community of commercial fishers is usually set along a coastline, be it a great lake or an ocean.

The setting of a community helps define the community because people select a community for what it has to offer them within the context of its landscape. The setting therefore helps create many characteristics that are unique to the community. By the same token, the values and development practices of a community alter the characteristics of its surrounding environment.

In addition to the surrounding environment, the built environment within a community is part of its setting and therefore its identity. This includes the buildings, zoning, design of transportation systems, and the allowance of natural occurrences within the structured setting.

In turn, a community's worldview defines its collective values, which determine how it treats its surrounding landscape. As the landscape is altered through wise use or through abuse, so are the community's options altered in like measure. A community and its landscape are thus engaged in a mutual, self-reinforcing feedback loop as the means by which their processes reinforce themselves and one another.

Each community has physical, cultural, and political qualities that make it unique and more or less flexible. The degree of flexibility of these attributes in a community is important because sustainable systems must be ever flexible, adaptable, and creative.[43] The process of sustainable development must therefore remain flexible because what works in one community may not work in another or may work for different reasons.

Beyond this, the power of sustainable development comes from the local people as they move forward through a process of growing self-realization, self-definition, and self-determination. Such personal growth opens the community to its own evolution within the context of the people's sense of place, as opposed to coercive pressures applied from the outside.

Sustainable development encompasses any process that helps people meet their requirements, from self-worth to food on the table, simultaneously creating a more ecologically and culturally sustainable and just society for the current generation and those that follow. Due to its flexibility and openness, it is perhaps more capable than other forms of development in creating such outcomes because it integrates the requirements of a local community with those of the immediate environment and surrounding landscape while instilling a relative balance between the local community and the larger world of which it is a part.

Between People in the Present and Those of the Future

We, as well as world political leaders of all persuasions, must now address a moral question: Do those living today owe anything to the future? If our answer is no, then we surely are on course because we are consuming resources and polluting Earth as if there were no tomorrow. If, on the other hand, the answer is "Yes, we have an obligation to the future," then we must determine what and how much we owe because our present, nonsustainable course is rapidly destroying the environmental options for generations to come. Meeting this obligation will require a renewed sense of morality—to be other-centered in doing unto those to come as we wish those before us had done unto us.[44]

To change anything, we must reach beyond where we are, beyond where we feel safe. We must dare to move ahead, even if we do not fully understand where we are going, because we will never have perfect knowledge. We must ask innovative, other-centered, future-oriented questions to make necessary changes for the better.

True progress toward an ecologically sound environment and an equitable world society will be expensive in both money and effort. The longer we wait, however, the more disastrous become the environmental conditions and the more expensive and difficult become the necessary social changes.

No biological shortcuts, technological quick fixes, or political promises can mend what is broken. Dramatic, fundamental change, both frightening and painful, is necessary if we are really committed to the children of the world, present and future. It is not a physical question of whether we can change but rather of humility: Will we change or will we insist on the status quo? If we are to approach social-environmental sustainability while there is still time, we must commit ourselves to do so and then get on with it. The outcome of such a commitment rests on two questions: (1) When is enough, enough? (2) To what extent are the consequences of our decisions repairable if they prove to be detrimental to social-environmental sustainability?

The Questions We Ask

Old questions and the old answers have led us into the mess we are in today and are leading us toward an even greater mess tomorrow. We must therefore understand that the answer to a problem is only as good as the question and the means used to derive an answer.

Before we can obtain fundamentally new answers, we must be willing to risk asking fundamentally new questions. This means that we must look long and hard at where we are headed with respect to the quality of our environment and to the legacy we are leaving the children of the world.

The questions we ask guide the evolution of humanity and its society, and it is the questions we ask, not the answers we derive, that determine the options we bequeath to future generations. Answers are fleeting, here today and gone tomorrow, but questions may be valid for a century or more. Questions are flexible and open ended, whereas answers are rigid, illusionary cul-de-sacs. The future, therefore, is a question to be defined by questions.

When Is Enough, Enough?

When is enough of something enough? This seems like a simple question. When you have eaten your fill, for example, you quit eating because you know you have had enough, for the moment at least. But, what about "enoughness" in the material sense, other than being immediately satiated with food?

Recall the previous discussion of the forest of Čergov, in Slovakia, that had been logged with horses in a biologically sustainable manner for centuries. It had been logged sustainably because the people of the small mountain villages who did the logging had enough of what they needed to satisfy them.

Their vibrant, centuries-old communities had been in harmony with the landscape. They had enough forest to produce quality timber to keep the people employed for generations. They had clean water and clean air. They had fertile soil for farming, which was combined in the community with the logging of timber and the milling of lumber, which was sold in the cities. And, they passed an intact cultural heritage from generation to generation.

But then, the Communists who controlled the Slovakian government decided that they did not have enough fast enough. So, they destabilized the centuries-old social-environmental harmony of the mountain villages from the outside. To satisfy their wants, they introduced nonsustainable technology in the name of modern forestry:

1. Chain saws replaced handheld crosscut saws.
2. Caterpillar tractors and logging trucks replaced horses.
3. Clear-cut logging replaced careful, single-tree selection logging.
4. Imported loggers replaced local loggers.

5. Distant sawmills replaced local mills.
6. Local jobs were replaced by unemployment or long commutes to distant cities.
7. The value of time-honored skills was replaced by a sense of present worthlessness.
8. The continuity of the centuries-old cultural heritage was replaced by an instantaneous loss of cultural identity.
9. Economically designed monocultural plantations replaced the diverse forest.
10. Erosion replaced soil stability and formation.
11. Flooding replaced nature's water infiltration and storage.
12. All this happened within less than a decade because enoughness was replaced by greed.

What about us in the United States? As an admittedly—even *proudly*—materialistic society, we tend to ask ourselves over and over such questions as

1. What do I want?
2. How much do I want?
3. How much can I get?
4. How fast can I get it?
5. How can I get it for little or nothing?

Society's present, collective wants, desires, needs, and demands are rapidly outstripping, if they have not already, the world's capacity to supply them and do not even take into account the wants, desires, needs, and demands of such rapidly industrializing nations as China and India, to say nothing of the coming generations. They also do not take into account the incredibly insane waste of global resources in war—any war for any reason.

"Necessity" is a very different proposition from the collective want-desire-need-demand syndrome. If we are wise enough to curb our appetites and to embrace the concept of necessity instead of want, Earth still has, we believe, enough resources to nurture us while we relearn how to nurture Earth. Nurturing Earth brings us to the concept of reversibility.

Are the Consequences of Our Decisions Reversible?

Although sustainability potentially requires absolute reversibility in our decisions and subsequent actions if they prove ecologically or socially non-sustainable, the consequences of our decisions are totally irreversible, but the consequences may be repairable in varying degrees in space and time. To understand this, let us consider three examples: a ditch, a dam, and soil.

A Ditch

If one wants to drain a swamp, one can dig a series of ditches, which lead water away from the swamp, and effectively lower the level of the water table. In so doing, the habitat changes from a swamp to a meadow, a field, or something else, but only as long as the ditches remain functional.

If, however, one wants to reclaim the swamp, one must refill the ditches. Although the water table will rise accordingly in the short term, it will take much longer for the life of the swamp to return. And, because the swamp habitat has been altered, it will never be as it would have been had the swamp not been drained. For all intents and purposes, however, this is an example of our illusion of reversibility, which you will better understand in Chapter 3 when you read "Principle 9: All Relationships Are Irreversible."

A Dam

Although dams can afford considerable economic and social benefits, their placement and construction must be grounded in sufficient knowledge of the river and its catchment basin to account for long-term ecological consequences. Dams are highly individualistic, and the same physical circumstance may elicit dramatically different responses. The effects of a dam in time and space not only can be environmentally gargantuan but also can become initially apparent after the elapse of years or decades.

While I (Chris) was working as a vertebrate zoologist with a scientific expedition in Egypt in 1963 and 1964, a representative of the Egyptian Ministry of Agriculture spent time with us as we worked just north of the Sudanese border along the Nile. One day, three of us from the expedition tried to help this man understand that building the High Aswan Dam across the Nile River was an ecological mistake. He could not see beyond the storage of water for the generation of electricity, however, which was the official position of the government.

We explained to the government representative that building the dam would increase the geographical distribution of the snails that carry the debilitating disease schistosomiasis, caused by a tiny blood fluke, from below the "low" Aswan Dam (built by the British in the early 1930s at the town of Aswan) south to at least Khartoum in the Sudan, several hundred miles (kilometers) above the new, yet-to-be-completed dam. At that time, it was still safe to swim above the Aswan Dam, where the water was too swift and too cold for the snails to live, but it was not safe to swim, or even catch frogs, in the water below the dam, where the snails already lived.

We told him that the Nile above the high dam would fill with silt, which would starve the Nile Delta of its annual supply of nutrient-rich sediment and affect farming in a deleterious way. We also told him that the dam could easily become a military target for the Israelis, as German dams were targets for the British during World War II.

The engineers building the High Aswan Dam had intended only to store more water and to produce electricity, which they did. However, deprived of the nutrient-rich silt of the Nile's annual floodwaters, the population of sardines off the coast of the Nile Delta in the Mediterranean diminished by 97 percent. In addition, the rich delta, which had been growing in size for thousands of years, is now being rapidly eroded by the Mediterranean because the Nile is no longer depositing silt at its mouth.

Moreover, the Mediterranean is starving for freshwater because human activities, such as the resulting Aswan High Dam, cut off most of the annual flow of the Nile River, which is now used for irrigation and no longer enters the sea. In addition, evaporation from the surface of the Mediterranean is increasing due to global warming. Consequently, a larger amount of freshwater is being lost to human activities and evaporation than is being replaced by rainfall and the inflowing of freshwater rivers. All of which means the Mediterranean is becoming more saline, and that salinity is being modified at the Strait of Gibraltar, where the waters of the Atlantic and Mediterranean mix. Barring a significant change in the regional circulation of the atmosphere, two human-caused losses of freshwater from the Mediterranean (the Aswan Dam and global warming) will cause the salinity to increase for some time because of the influence burning fossil fuels has on global warming.[45]

Until the High Aswan Dam was built, the annual sediment-laden waters of the Nile added a millimeter (a little less than a 16th of an inch) of nutrient-rich silt to the farms along the river each year. Now that the dam has stopped the floods, the silt not only collects upriver from the dam, thus diminishing its water-holding capacity, but also is no longer being deposited on the riverside farms, thus decreasing their fertility. Also, schistosomiasis has indeed spread southward to the Sudan.

And, a people, the Nubians, were displaced and their culture thereby destroyed. The Nubians I got to know were tall, straight, slender, and very black with fine features and lived many miles (kilometers) south of Aswan on small farms sandwiched between the east bank of the Nile and the Eastern Desert. Their village was neat and clean, and plates, with designs around their borders, were embedded into the outer mud coating of the doorway of each house, thus decorating them. The people had a wonderful sense of humor, were quick to laugh, and seemed genuinely pleased that I delighted in playing with their children and vice versa. For their part, the children seemed to have a good sense of self and of each day as an adventure to be lived to the fullest.

But, the High Aswan Dam changed all that. The Nubians were moved inland from the bank of the Nile, whose quiet flowing waters and silent guardian desert had been a part of their lives seemingly forever. In place of their freely spaced, cool, airy, self-designed, and self-constructed homes, they were put into government-built, look-alike, minimum-quality housing, where many of them could not adjust and simply died.

There is yet another consequence of the Aswan High Dam, one I would never have thought of, even though I had studied the mammals along the Nile. The Nile annually flooded the many nooks, crannies, and caves along its edge, killing the rats whose fleas carry bubonic plague. Because the floods no longer occur, the rat population has soared, and bubonic plague is once again a potential threat.

I learned about this unexpected consequence of the Aswan High Dam from Dr. Wulf Killmann of the Deutsche Gesellschaft für Technische Zusammerarbeit, whom I met in Malaysia. As we visited about the effects of dams on rivers and oceans, I told him about my experience in Egypt. Dr. Killmann then told me that he had been part of a project to figure out how to control the ever-growing population of rats, which has become a serious health problem.

What would happen if the High Aswan Dam were to be removed? While the floods would once again begin fulfilling their many biophysical roles, there would be an immediate problem of how to deal with all the silt trapped behind the dam. Then, there is the question of what to do with all the steel, concrete, and other materials of which the dam is built. What would happen to all the economic investments and technological developments that have over the years sprung into existence because of the dam? How would the Egyptian people replace the social benefits engendered by the dam? And, there is at least one benefit: greater electrical capacity for the country.

Even if all conceivable questions could be answered and most of the effects could be to some extent repaired, there is at least one that is final. The Nubian culture, in which I found such beauty and joy, would still be extinct. So, the question is: How "reversible" in reality are the effects of the Aswan High Dam?

Soil

Because traditional economics in our linear capitalistic system does not acknowledge intrinsic ecological value as "real" value, we guide the use and attempted management of our natural resources only by the cost-benefit analysis of their potential economic value when converted to something else, such as from trees to boards, forage to red meat, water to hydroelectric power, and so on. This means that the only value economists can see is short-term specialization, a view that kills the very soils on which we depend for sustenance.

Many cultures have emphasized the trusteeship of the soil through religion and philosophy. Abraham, in his covenant with God, was instructed: "Defile not therefore the land which ye shall inhabit, wherein I dwell."[46] Chinese philosopher Confucius saw in Earth's thin mantle the sustenance of all life and the minerals treasured by human society. And a century later, the Greek thinker Aristotle viewed the soil as the central mixing pot of air, fire, and water that formed all things.

In spite of the durability of such beliefs, many cannot grasp them because they are intangible. The invisibility of the soil is founded in the notion that it is as common as air and therefore is taken for granted, as is air. To many people, the soil seems "invisible." But, when we think about it, we realize that humanity, indeed human society, is somehow tied to the soil for reasons beyond measurable materialistic wealth.

Even though we can justify soil protection economically, our ultimate connection with it escapes many people. One problem is that traditional economics deals with short-term tangible commodities, such as fast-growing row crops, rather than with long-term intangible values, such as the future prosperity of our children. But, when we recognize that land, labor, and capital are finite, and that every system has a carrying capacity, one that depends on natural or artificial support, the traditional, linear economic system becomes more like a cyclic biological system.

In the late eighteenth century, Thomas Malthus proposed that the human population would grow faster than the soil's ability to sustain it, but agronomic advances in this century led many shortsighted leaders to dismiss this idea as simplistic and overly pessimistic. Today, however, Malthusian theory seems prophetic when one considers the air pollution that poisons the soil, overgrazing by livestock and the growing desertification, global deforestation and the loss of the soil's protective cover of vegetation and its vitality, and the recurring famines.

Those who analyze the soil by means of traditional, linear economic analyses weigh the net worth of protecting the soil only in terms of the expected short-term revenues from future harvests and ignore the fact that it is the fertile integrity of the soil that produces the yields. In short, the protection of the soil is seen as a cost with no benefit because the standard method for computing "soil expectation values" and economically optimal crop rotations commonly assumes that the productivity of the soil will either remain constant or increase—but never decline.

In reality, however, reducing the productivity of the soil on marginal sites, the logic of which is both shortsighted and flawed, can push the expected present net worth of subsequent harvests below zero. Given that reasoning, it is not surprising that those who attempt to *manage* the land seldom see protection of the soil's productivity as cost effective. If we could predict the real effects of *management* practices on long-term economic yields, we might have a different view of the invisible costs associated with poor care of the soil.

Returning to the Question: What Is Sustainable Development?

The problems in our communities can neither be isolated nor understood without first understanding their context. This means that one must understand the various parts of a community and their interactions before one can understand why something is the way it is, which raises the question: What is "sustainable development?"

In developing this understanding, process has primacy over the parts because process directs the outcome, the function of each part. A part does not control the process, although it may influence it.

In placing development within a new context of conscious choice, the answers to such questions as: What do we mean by development? What is underdeveloped? What is poverty? will be very different than they are today. If a lifestyle promotes sustainability through conscious choice, conscious simplicity, and self-provisioning and recognizes the relationships between one's own sustenance and the livelihood of one's immediate surroundings (one's fidelity to a sense of place) in relationship to the larger world, that life is not necessarily perceived as one of poverty. This leaves the way open to change the indicators of development.

Progress, therefore, would be any action that moves a person, community, culture, or society toward social-environmental sustainability. For society to progress, decisions must be made that recognize and respect the requirements and rights of future generations, as well as the requirements and intrinsic value of all species and Earth's biophysical carrying capacity with respect to its human population. (Carrying capacity is the number of individuals who can live in and use a particular landscape without impairing its ability to function in an ecologically sustainable way.) This position is very different from our blind faith in technology and material acquisition, which we think of as progress and thus as *development*.

Sustainable development is thus about the notions of *enoughness* and what can be *repaired*. Here, the operative questions are: When is enough enough? and, If we err in our decision, is the outcome repairable? Such questions are necessary to promote a change in the nature of social-environmental decisions. What is needed to resolve our social-environmental problems goes beyond environmentally safe commodity production and technology.

Instead of the current tinkering, which deals solely with the symptoms of our social-environmental malaise, problems must be solved at their source—worldview assumptions and values—because these drive our decisions, policies, and plans. Sustainable development therefore questions the very purpose of society and our participation with our home planet and demands social-environmental justice, which challenges the very heart of our perceived relationship with nature and one another, present and future.

We, as planetary citizens, must learn to think at least 10 generations ahead when making decisions because the great and only gift we have to give those who follow are potential choices and some things of value from which to choose. Today's decisions become tomorrow's consequences—a notion that highlights the word *responsibility*.

Responsibility is a double-edged sword in that our responsibility, our moral obligation, is to choose carefully today so that the generations to come can respond viably to the circumstances we have created for their time of choice. Intelligent decisions on our part are possible only when we both recognize and accept the intrinsic value of nature as a living organism rather than accepting nature only as a collective resource (host) from whose body we extract (parasitize) a variety of commodities as the life's blood of our dysfunctional, linear economic system.

Development must be flexible and open to community definition because the values promoted must meet various needs and situations in space and time while safeguarding sustainability. The process of valuation embodied in sustainable development addresses social-environmental justice in recognizing the necessity of equal access to resources as well as equal distribution of goods and services while protecting the long-term biophysical sustainability of the system that produces them.

Sustainable development also addresses the need to promote education and feelings of self-worth in people, allowing them to act as catalysts in the process of change, whether in their own lives or in the life of society. For change to be a creative process, each person must respect every other person as well as the intrinsic value of his or her environment.

Finally, the valuation/decision process, through which sustainable development works, must flow within and promote the democratic frame of reference because democracy only works when it is actually practiced. In this sense, most of the change must be directed by the people from the bottom up—the "grass roots" of the local community for it to be what we discuss further as a local living economy.

Summary

Chapter 1 examines the interconnectedness among various aspects of nature and society as they form an integrated, interactive whole. The primary thrust of this discourse contends that we humans are not only inseparable from nature but also have a right to be here and participate with nature. That our mere existence changes the world is a given simply because we exist and use energy. The issue is how and why we alter our home planet because therein we, the present adults, have a choice. To all the generations who follow, however, we bequeath the consequences of our decisions and actions—and they have no alternative but to live with them as best they can.

With the foregoing in mind, Chapter 2 addresses the human components that make a sustainable community, such as a sense of place, history, and trust.

Endnotes

1. Arnold J. Toynbee. *A Study of History*, Volumes 1–6 (abridgement by D. C. Somervell). Oxford University Press, New York, 1987.
2. T. N. Narasimhan. "Limitations of Science and Adapting to Nature." *Environmental Research Letters* 2 (July–September 2007) 034003, doi:10.1088/1748-9326/2/3/034003 (accessed October 1, 2011).
3. (1) Rod Swenson. "Emergent Evolution and the Global Attractor: The Evolutionary Epistemology of Entropy Production Maximization." In: *Proceedings of the 33rd Annual Meeting of the International Society for the Systems Sciences*, P. Leddington (ed.), 33(3), 46–53, 1989; and (2) Rod Swenson. "Order, Evolution, and Natural Law: Fundamental Relations in Complex System Theory." In: *Cybernetics and Applied Systems*, C. Negoita (ed.), 125–148. New York: Dekker, 1991.
4. Rod Swenson and Michael T. Turvey. "Thermodynamic Reasons for Perception–Action Cycles." *Ecological Psychology* 3 (1991):317–348
5. Rod Swenson. "Spontaneous Order, Autocatakinetic Closure, and the Development of Space-Time." *Annals New York Academy of Sciences* 901 (2000):311–319.
6. Narasimhan, "Limitations of Science."
7. Norman Jacob. "Towards a Theory of Sustainability." *Trumpeter* 6 (1989):93–97.
8. John J. Magnuson. "Long-Term Ecological Research and the Invisible Present." *BioScience* 40 (1990):495–501.
9. The foregoing discussion of ecosystem function is based on Chris Maser. *Earth in Our Care: Ecology, Economy, and Sustainability*. Rutgers University Press, New Brunswick, NJ, 2009.
10. The discussion of Easter Island is based on (1) Michael Kiefer. "Fall of the Garden of Eden," *International Wildlife* July–August (1989):38–43; (2) Terry L. Hunt and Carl P. Lipo. "Late Colonization of Easter Island." *Science* 311 (2006):1603–1606; and (3) Terry L. Hunt. "Rethinking Easter Island's Ecological Catastrophe." *Journal of Archaeological Science* 34 (2007):485–502.
11. Fritz M. Heichelheim. "The Effects of Classical Antiquity on the Land. In: *Man's Role in Changing the Face of the Earth*, W. L. Thomas (ed.), 165–182. University of Chicago Press, 1956.
12. (1) Robert Cullen. "The True Cost of Coal." *Atlantic Monthly* December (1993):38, 40, 48–50, 51; (2) Sid Perkins. "What Goes Up." *Science News* 172 (2007):152–153, 156; (3) Robert Vautard, Pascal Yiou, and Geert Jan van Oldenborgh. "Decline of Fog, Mist and Haze in Europe over the Past 30 Years." *Nature Geoscience* 2 (2009):115–119; (4) V. Ramanathan and G. Carmichael. "Global and Regional Climate Changes Due to Black Carbon." *Nature Geoscience* 1 (2008):221–227; (5) Drew T. Schindell, Greg Faluvegi, Dorothy M. Koch, and others. "Improved Attribution of Climate Forcing to Emissions." *Science* 326 (2009):716–718; (6) C. Warneke, R. Bahreini, J. Brioude, and others. "Biomass Burning in Siberia and Kazakhstan as an Important Source for Haze over the Alaskan Arctic in April 2008." *Geophysical Research Letters* 36 (2009) L02813, doi:10.1029/2008GL036194 (accessed September 24, 2012); and (7) Michael Fromm, Daniel T. Lindsey, René Servranckx, and others. "The Untold Story of Pyrocumulonimbus." *Bulletin of the American Meteorological Society* 91 (2010):1193–1209.

13. The preceding three paragraphs are based on (1) Charles E. Little. "Report from Lucy's Woods." *American Forests* March/April (1992):25–27, 68–69; (2) R. J. Esher, D. H. Marx, S. J. Ursic, and others. "Simulated Acid Rain Effects on Fine Roots, Ectomycorrhizae, Microorganisms, and Invertebrates in Pine Forests of the Southern United States." *Water, Air, and Soil Pollution* 61 (1992):269–278; and (3) Walter C. Shortle and Ernest A. Bondietti. "Timing, Magnitude, and Impact of Acidic Deposition on Sensitive Forest Sites." *Water, Air, and Soil Pollution* 61 (1992):253–267
14. (1) Paul Recer. "Old Pesticides Spread across Globe." *Corvallis Gazette-Times* (Corvallis, OR) October 5, 1995; and (2) Sustaining Lake Superior. "Toxaphene." 2011. http://www.sustaininglakesuperior.com/3/post/2011/05/toxaphene.html (accessed November 14, 2012).
15. The preceding two paragraphs are based on Elaine R. Ingham. "Organisms in the Soil: The Functions of Bacteria, Fungi, Protozoa, Nematodes, and Arthropods." *Natural Resource News* 5 (1995):10–12, 16–17.
16. The preceding four paragraphs are based in part on: Chris Maser. "The Humble Ditch." *Resurgence* 172 (1995):38–40.
17. The foregoing discussion of forests is based on (1) Chris Maser. *Our Forest Legacy: Today's Decisions, Tomorrow's Consequences*. Maisonneuve Press, Washington, DC, 2005; and (2) Chris Maser, Andrew W. Claridge, and James M. Trappe. *Trees, Truffles, and Beasts: How Forests Function*. Rutgers University Press, New Brunswick, NJ, 2008.
18. Jeremy Rifkin. *Biosphere Politics: A Cultural Odyssey from the Middle Ages to the New Age*. HarperSanFrancisco, San Francisco, 1991, 388 pp.
19. The preceding two paragraphs are based on Chris Maser. *Sustainable Forestry: Philosophy, Science, and Economics*. St. Lucie Press, Delray Beach, FL, 1994.
20. Noam Chomsky. "How Free Is the Free Market?" *Resurgence* 173 (1995):6–9.
21. (1) Chris Maser, Russ Beaton, and Kevin Smith. *Setting the Stage for Sustainability: A Citizen's Handbook*. Lewis, Boca Raton, FL, 1998; and (2) Russ Beaton and Chris Maser. *Economics and Ecology: United for a Sustainable World*. CRC Press, Boca Raton, FL, 2012.
22. Sarita Chawla and John Renesch (eds.). *Learning Organizations: Developing Cultures for Tomorrow's Workplace*. Productivity Press, New York, 1995.
23. Beaton and Maser, *Economics and Ecology*.
24. Duncan M. Taylor. "Disagreeing on the Basics." *Alternatives* 18 (1992):26–33.
25. Beaton and Maser, *Economics and Ecology*.
26. Toynbee, *A Study of History*.
27. The preceding two paragraphs are based on Monica G. Turner, Eugene P. Odum, Robert Costanza, and Thomas M. Springer. "Market and Nonmarket Values of the Georgia Landscape." *Environmental Management* 12 (1988):209–217.
28. Elizabeth Ann R. Bird. "The Social Construction of Nature: Theoretical Approaches to the History of Environmental Problems." *Environmental Review* 11 (1987):255–264.
29. Wendell Berry. "Word and Flesh." *Whole Earth Review* Spring (1990):68–71.
30. Rifkin, *Biosphere Politics*.
31. The foregoing three paragraphs are based on Monica G. Turner. "Landscape Ecology: The Effect of Pattern on Process." *Annual Review of Ecological Systems* 20 (1989):171–197.
32. Turner et al., "Market and Nonmarket Values."

33. Chris Maser. "Do We Owe Anything to the Future?" In: *Multiple Use and Sustained Yield: Changing Philosophies for Federal Land Management?* Proceedings and summary of a workshop convened on March 5 and 6, 1992, Washington, DC. Congressional Research Service, Library of Congress. Committee Print No. 11 (1992):195–213. U.S. Government Printing Office, Washington, DC.
34. Kirk Talbott. *Central Africa's Forests: The Second Greatest Forest System on Earth*. World Resources Institute's Center for International Development and Environment, Washington, DC, 1993, 12 pp.
35. Walt Bresette. "The 7th Generation Amendment." 1996. http://www.greens.org/s-r/09/09-19.html (accessed November 18, 2012).
36. The preceding discussion of conservatives is based on David Orr. "Conservatives against Conservation." *Resurgence* 172 (1995):15–17.
37. Edmund Burke. http://quotes.dictionary.com/men_are_qualified_for_civil_liberty_in_exact (accessed September 15, 2012).
38. The foregoing discussion of blunders is based on Donella Meadows. "Seven Blunders." *Resurgence* 172 (1995):13.
39. Ibid.
40. Ibid.
41. Terry Tempest Williams. *The Open Space of Democracy*. Orion Society, Great Barrington, MA, 2004, 107 pp.
42. Robert Constanza and Herman E. Daly. "Toward an Ecological Economics." *Ecological Modelling* 38 (1987):1–7.
43. Robert Rodale. "Big New Ideas—Where Are They Today?" Unpublished speech given at the Third National Science, Technology, Society Conference, February 5–7, 1988, Arlington, VA.
44. Maser, "Do We Owe Anything."
45. The foregoing discussion of the Aswan High Dam is based on (1) C. J. George. "The Role of the Aswan Dam in Changing Fisheries of the South-Western Mediterranean." In: *The Careless Technology*, M. T. Farvar and J. P. Milton (eds.). Natural History Press, New York, 1972; 159–178; (2) James P. M. Syvitski, Charles J. Vörösmarty, Albert J. Kettner, and Pamela Green. "Impact of Humans on the Flux of Terrestrial Sediment to the Global Coastal Ocean." *Science* 308 (2005):376–380; (3) Sid Perkins. "Muddy Waters." *Science News* 167 (2005):328–329; and (4) R. G. Johnson. "Climate Control Requires a Dam at the Strait of Gibraltar." *EOS, Transactions, American Geophysical Union* 78 (1997):277–281.
46. *Holy Bible*, Authorized King James Version. World Bible, Iowa Falls, IA, Numbers 35:34.

2

True Community Is Founded on a Sense of Place, History, and Trust

Introduction

Community is rooted in a sense of place, where people live in a reciprocal relationship with their landscape. As such, a community is not simply a static place within a static landscape but rather a lively, ever-changing, interactive, and interdependent system of biophysical relationships—both social and environmental. Because a community is a self-organizing system, it does not simply incorporate information, but through its activities changes its environment as well.

In a sense, the distribution of communities across a landscape is analogous, geographically, to nations across a continent. And, "geography is nothing less than the social, environmental, and political implications of humanity's interaction with the landscape," says author-editor Robert D. Kaplan.[1]

The United States, writes Kaplan, was founded in a state of nature. Our form of self-government, we are reminded by such historians as Fredrick Jackson Turner and Daniel J Boorstin, "evolved not from ideology, as in Europe, but from encounters with the landscape by the early settlers."[2] Thus, as a community alters the landscape, so the landscape, in reaction, alters the community.

Reciprocity is the self-reinforcing feedback loop that either extends sustainability to, or withholds it from, a community and its landscape. Thus, a person who is mindful and feels a deep sense of place can bring about change that is beneficial to the social-environmental sustainability of his or her community. But, if, after some time, a person is told by his or her employer that a transfer is likely within 2 years, that person tends to go from a resident at heart, with a sense of place, to a transient who is reluctant to invest any more money or time in his or her home or in the community. As more and more people are put in such a position, changes that take place in a community tend to tear at its fabric, which in a collective, historical sense is part of the collateral damage.

Community History

History is a reflection of how we see ourselves and thus goes to the very root of how we attach value to things. Our vision of the past is shaped by, and in turn shapes, our understanding of the present—those complex and comprehensive images we carry in our heads with which we decide what is true or false.[3]

A community's history must therefore be passed from one generation to the next if the community is to know itself throughout the passage of time. In this sense, it is ironic that during the greatest era of prosperity in the United States, the decades following World War II, only the cheapest possible buildings were constructed, including civic buildings. To understand what we mean, compare any richly designed post office or city hall built at the turn of this century with its modern, unimaginative, concrete-box counterpart.

When the United States was a far less wealthy nation (by monetary standards), things were built to endure because it would have seemed immoral, if not insane, in our great-grandparents' day to throw away hard-earned money and honest labor on something guaranteed to disintegrate within 30 years.

The buildings erected in those earlier days paid homage to history in their design, including elegant solutions to the age-old problems posed by the cycles of weather and light. They paid respect to the future because they were consciously built to endure beyond the lifetimes of the people who constructed them—a continuum of past, present, and future or "chronological connectivity." Chronological connectivity puts us in touch with the ages and connects us with a sense of eternity, indicating that we are somehow part of an organism that is significantly larger than ourselves.

The notion of chronological connectivity suggests that the large organism we help to compose—our community—even cares about us, which in turn nurtures self-respect and respect for those who will follow us, just as those who preceded us respected those who were to follow. In this case, connectivity with the memories of the past and the horizon of the future lead us in the direction of social-environmental sustainability.

If the connective continuity of a community is disrupted, the erosion of community identity is likely to follow. The extinction of identity occurs when the community begins to view its landscape not as an inseparable extension of itself but rather as a separate commodity and, all too commonly, a commodity to be exploited for immediate financial gain. When this happens, community is corrupted or even destroyed from within because trust is withdrawn in the face of growing numbers who are either transient or are absentee landowners.

Our graceless built environment, with its many structures designed to disintegrate within a few decades, suggests a rejection of and disrespect for both past and future. This consciously built-in obsolescence is euphemistically termed "design life," which may last 50 years. Since today's buildings are expected to serve only our era, we seem unwilling to expend money or effort for either their beauty or their service to future generations.[4]

While elements of "green building" are gradually making their way into the realm of "best practices," the building industry, generally speaking, does not seem to invest, as a matter of practice, in the development of elegant solutions to the problems created by the cycles of weather and light. Rather, it continues to rely heavily and consistently on such technology as electricity, central heating, and air conditioning. Thus, many new office buildings have windows that cannot be opened or virtually no windows at all. This process of disconnecting from the time continuum of the past, through the present, into the future and from the cycles of weather and light diminishes us spiritually, impoverishes us socially, and destroys the time-honored cultural patterns we call community.

While this still appears to be the dominant pattern of development, the forces of the green building and sustainable built environment are gaining traction. The Living Building Challenge, for example, developed by green building leader Jason McLennen, calls for new structures to produce all of their own energy and use only water that falls on site. Builders choosing to follow this standard must use materials from sustainable sources and avoid toxic materials, including asbestos, mercury, and polyvinyl chloride (PVC). "They must build on previously developed, sites and meet measurements of livability, social equity, and beauty."[5]

The Living Building Challenge's seven performance standards, which McLennan calls "petals," evoking the elegance and efficiency of a flower, "are far more demanding than even the highest level of Leadership in Energy and Environmental Design, the most widespread green building certification." And yet, it is clear that the vast majority of developers will go about their business as usual.[6]

As new ideas, such as the Living Building Challenge, are embedded in the larger context of the Living Future Institute and gain a following, conflict between sets of values in communities must be addressed. Many communities are in transition between sets of values, which must be carefully assessed in terms of their overall sustainability before change can be envisioned. This applies to both attitudes and land use practices. As authors James and Roberta Swan point out: "There is a need to find a common language and conceptual framework to promote mutual understanding about the power of place. It is easy to feed the fires of conflict … the more difficult task is to build bridges of respect and cooperation."[7]

Community can also be lost another way: when citizens become resistant to the costs of maintenance. We Americans seem eager to build but then begrudge providing the tax dollars necessary to maintain highways and schools, let alone downtowns, which we have effectively abandoned in many instances—a trend we have seen for decades in the proliferation of suburban shopping centers and malls, located conveniently off freeways or adjacent to primary arterials, far from city centers.

One of the reasons some are reluctant to spend money on maintenance is a lack of long-term commitment to the community. Where one lives while earning a living may not hold sufficient value for them to consider retiring there. Thus, while they are committed to working in and living as a part of the town, they plan to move somewhere else to spend the rest of their lives on retirement.

The United States has the highest rate of transience on the planet, with an average residency of 5.9 years, according to 2004 U.S. Census data.[8] With this as a reality, why would a person destined to live for only 6 years in a community make a significant investment in maintaining a place he or she is most likely planning to leave?

The result is that we lose what we neglect, be it a house, a street, or a downtown. Sustainable communities are not made to be disposable; they are not designed in terms of planned obsolescence. Tithing a portion of one's time to do something to improve the quality of one's community is the beginning of recognizing the difference between real wealth and money. Conventional money knows no loyalty to a particular place, a local community, or even a nation, and so it flows toward a global economy in which traditional social bonds give way to a rootless quest for the highest monetary return. The real price we pay for money, the real cost, many times is the hold it has on our sense of what is possible—the prison it builds around our imagination.

Money versus Wealth

Bernard Lietaer, of the Center for Sustainable Resources at the University of California at Berkeley, attests that, "Money is like an iron ring we've put through our noses. We've forgotten that we designed it, and it's now leading us around. I think it's time to figure out where *we* want to go—in my opinion toward sustainability—and then design a money system to get us there."[9] That said, the way we spend money makes our private beliefs public declarations.

Although textbooks on economics claim that people and corporations are competing for resources and markets, they are in reality competing for money, and in so doing are using resources and markets. According to author Caroline Myss:

> Money, like energy, is a neutral substance that takes its direction from the intention of the individual. A more fascinating aspect of money is the fact that it can weave itself into the human psyche as a substitute for the life force. When people equate money with their life energy—an often-unconscious substitution—the consequences are usually negative, for every dollar a person spends is also an unconscious expenditure of energy.[10]

"Modern money," explains economist and author David Korten, "is only a number on a piece of paper or an electronic trace in a computer that by social convention gives its holder *a claim* on real wealth," which he goes on to say has concrete value in meeting the necessities of and fulfilling our desire for a quality life.[11] But, in our confusion over the true nature of *real wealth* (if we have thought about it at all), we chase after dollars and neglect those things that actually sustain a life of quality—both spiritual *and* material. Money has only extrinsic value, only the potential to be converted into something else that may have the intrinsic value of real wealth.

A fixation with the "money chase" makes it difficult to express the basic difference between money and wealth through the symbology of language. It is not difficult to see that real wealth comes from the biophysical integrity of the diverse ecosystems on which we rely. They not only are life support systems but also the basis for social-environmental sustainability. Real wealth means human equality and dignity, meaningful work, having a good home and adequate food, a good education, and so on. In a healthy economy, money serves the people in helping to create and protect real wealth but is neither the dominant value nor the sole—or even the main—medium of exchange.

One of the most important indicators of economic health is social-environmental sustainability, which means not only quality interpersonal relationships but also quality, reciprocal relationships between people and their environment. A truly viable economy is based on love and reciprocity, with people doing kind and useful things for one another with no expectation of financial gain. Such mutual caring is the soft social capital that both creates and maintains the fabric of trust, which in turn is the glue of functional families, communities, and societies.

Pathology and its subsequent dysfunction enter the economic system when money, derived as a convenient means of exchange, becomes the factor that defines the purpose of life for individuals and their communities. Illustrative of this point is the following: The majority of Americans in a November 28 to December 1, 2011, poll said they would need an annual income of $150,000 to feel rich. And, men had a higher median threshold, $150,000, than women—$100,000. Moreover, as people earn more, they typically want more. To wit, people earning $75,000 or more say they would need a median of $250,000 to feel rich.[12]

The dominance of money as the primary driver of society is also revealed in the increasing commercialization of human relationships. Not long ago, even in such rich, industrialized countries as the United States, at least half of the adult population—predominantly women—worked without a salary to create and maintain home and community, which are among the most fundamental values of a viable, functionally sustainable society and its social-environmental economy.

Today, financially supporting a household usually requires two adults holding two, and sometimes three or four, paying jobs between them, but at the expense of quality human relationships because, of necessity, they rarely see each other, and the care of children and the home is either neglected or hired out. In addition, the once-shared, mutual caring for community is now primarily in the hands of hired public employees—to the extent the public is willing to pay for it.[13]

The Value of Social Capital

As the soft social capital of mutual caring dwindles and the resulting quality of family life withers, a community's members typically become increasingly alienated, apathetic, or competitive. As human relationships become increasingly dysfunctional, a community's infrastructure crumbles into ever-greater disarray at an ever-increasing social-environmental cost to all generations. Today, more than ever, the wisdom of Henry Ford is sorely needed: "The highest use of capital is not to make more money, but to make money do more for the betterment of life."[14]

Consider by way of example the tiny hamlet of Roseto nestled into a wooded hillside in the Pocono Mountains of eastern Pennsylvania. Although the bricks and mortar that hold the ethnic Italian-American enclave together may not be distinguishable from its neighboring towns, Roseto is nonetheless distinct. Researchers in the 1960s found the citizens of Roseto, who numbered just over 1,600, were among the healthiest people in the United States.

Rosetan death from heart disease was half the national average, in addition to which Rosetans tended to live many years longer than their peers. They also exhibited greater resistance to mental illness and peptic ulcers than did other Americans. Their good health and longevity defied medical logic because they smoked and drank as much as other Americans, experienced as much stress, exercised as infrequently, and ate a high-fat diet in traditional Italian style. How could one account for such good health in a group with such traditionally unhealthy habits? This can be described in a word: community.

Dr. Stewart Wolf, MD, who discovered Roseto more than 30 years ago, found the tight-knit community that had evolved in Roseto fostered an atmosphere of friendship and mutual caring and support that buoyed and protected its residents from the stresses of everyday life and thereby protected them against heart disease. He named the health-giving benefit of living in a close community the "Roseto effect."[15]

After punching out on the time clocks at the local slate quarries and blouse factories, the people of Roseto returned home to close, extended families, where grandparents, parents, and children lived under one roof, each nurturing and supporting the other. They also invested themselves in the larger community by walking around the neighborhood after dinner to chat and joke with one another, and they joined and socialized formally at the many civic and community organizations and functions, from the Marconi Social Club to the Parent-Teacher Association. In addition, religion played an important role in this predominantly Catholic town.

In the 1970s, however, the close-knit community of Roseto began to unravel as the younger generation increasingly began to look for work elsewhere, attendance slipped in church, and the number of three-generational households diminished. The social conformity that had maintained narrow differences between the haves and have-nots disappeared as well, tearing the fabric and spirit of the community.

As the class differences between the haves and have-nots progressively widened, the social glue that had once held Roseto together dissolved; so did the now-legendary health benefit of living there. The very first fatal heart attack in anyone under 45 was recorded in 1971. Today, the health benefit of living in Roseto is history. The rate of heart disease mirrors the national average, and Rosetans live no longer than anyone else.[16]

From this scenario, it is clear that the fabric of communities needs to be rewoven. "In a society in which relationships are defined by love, generosity, and community," David Korten writes, "the importance of money in mediating personal exchange and allocating resources is likely to decline markedly."[17]

Reweaving the Social Fabric

Reweaving the social fabric is a complex endeavor. Reducing our dependence on money and restoring nonmonetary exchanges would, of course, aid the process. In recognition of this need, many are scaling down consumption, reducing their dependence on paid work while increasing their reliance on local products to meet basic necessities, thereby joining a movement that is strengthening the engagement of all persons in the productive life of family and community.

The sense of true community literally cannot extend itself beyond local place, mutual caring, and its history. Community, says Wendell Berry, "is an idea that can extend itself beyond the local, but it only does so metaphorically. The idea of a national or global community is meaningless apart from the realization of local communities."[18]

For a community to be founded in the first place and to be healthy and sustainable, it must rest on the bedrock of trust.

> A community does not come together by covenant, by a conscientious granting of trust. It exists by proximity, by neighborhood; it knows face to face, and it trusts as it knows. It learns, in the course of time and experience, what and who can be trusted. It knows that some of its members are untrustworthy, and it can be tolerant, because to know in this matter

> is to be safe. A community member can be trusted to be untrustworthy and so can be included. But if a community withholds trust, it withholds membership. If it cannot trust, it cannot exist.[19]

Although trust is the ability to rely on the integrity or to have faith in the character of a person, it cannot really be defined; it can only be lived in one's motives, thoughts, attitude, and behavior. Trust, therefore, must be based on truth. To make this point, Marc Luyckx, a member of the Forward Studies Unit of the European Commission, a think tank within the administration of the European Union in Brussels, Belgium, related the following story, which was told to him by the president of the World Business Academy. The story took place in northern Australia, where some reporters tape-recorded two policemen insulting an Aboriginal person. The reporters then printed the transcript on the front page of the newspaper.

The chief of police, an intelligent and open-minded person who had for some time been trying to reshape his police force, was to be interviewed about the incident on television. Hearing this, the minister of justice told the chief to fire the two policemen and warned him to be careful of what he said publicly or he would be fired also.

But, instead, on camera, the chief said: "I must tell the truth, and the truth is those two policemen are not an exception. The rest of the police could have done the same. But I will tell you more. The rest of the population of my region of Australia could have done the same as well. We are becoming racists."

There was an explosion in the press and on television, and the chief of police was fired the next day. The firing had not been approved by the Parliament, however, and the opposition party in the Parliament applauded him for a full minute for telling the truth in public. This action forced the government party to applaud as well, which resulted in the minister being fired and the chief being reinstated.[20] After truth has been spoken, there is not only room, but also permission, for spiritual growth and personal authenticity, which leads to trust—the essence of true community.

In sum, community is relationship, and meaningful relationship is the foundation of a healthy, sustainable community. In this connection, Ralph Waldo Emerson penned the following: "It is one of the most beautiful compensations of this life that no man can sincerely try to help another without helping himself."[21] As such, a resident community serves five purposes: (1) social participation, where and how people interact with one another to create the relationships necessary for a feeling of self-worth, safety, and shared values; (2) mutual aid, the services and support offered in times of individual or familial need; (3) economic production, distribution, and consumption, such as jobs, import and export of products, as well as the availability of such commodities as food and clothing in the local area; (4) socialization, which includes educating people about cultural values and acceptable norms; and (5) social control, the means to maintain those cultural values and acceptable norms.[22]

Community is a reminder that the scale of effective organization and action has always been small local groups. As anthropologist Margaret Mead said: "Never doubt that a small group of thoughtful, committed citizens can change the world; indeed it is the only thing that ever has."[23] Creating sustainable communities strengthens one's fidelity to a sense of place and is the best possible way to keep a community and its population intact because it raises the value of staying home. Top-down government cannot fulfill these things.

Grieving for Our Environmental/Social Losses

Abba Eban observed: "History teaches us that men [people] and nations behave wisely once they have exhausted all other alternatives."[24] The question with respect to land use planning is: Are we ready for the hard choices?

The basic challenge in making these hard choices stems from the fact that debates over policy are often flawed because the assumption is that the people involved share a common understanding of the issues being discussed. What is more, the assumptions, knowledge, and understanding of the issues are frequently uncertain and contested.[25]

Perhaps the clearest conclusion of all, amid the great complexity of today's world, is that small change(s) will not be sufficient to alter our present environmental degradation. Moreover, the change we must create is within ourselves—not in nature, science, or technology—and manifests in how we view the world around us.

Nevertheless, choice is the essence of sustainable community development. Yet, an increasing number of vested interests, such as residential developers, want to remove all emotion from land use decisions. They argue that it is not fair to developers when those people living in the area of a proposed housing subdivision keep delaying construction with legal appeals, despite the fact that the proposed developments meet legal requirements. They contend that the approval process, which includes the right of appeal if the public feels it is needed, costs developers too much money.

The effect of such thinking (1) steals choice and self-determining government from the people who live in the area of the proposed development; (2) gives preference to residential developers, with an increasing number seeming to be absentee, even from out of state; (3) forces local people to accept absentee interests; (4) limits and undermines the scope of a local community's potential vision for sustainable development within the context of its own landscape, especially for a desired future condition; and (5) curtails—or even eliminates—the ability of local people to actively mourn for the continuing loss of their quality of life and their sense of place as outside choices are forced on them, often by people who will not have to live with the consequences of the imposed actions.

Emotions, the force behind relationships, are based on personal and collective values—the heart and soul of community. Public debate, and the emotions it evokes, not only helps participants integrate the proposed changes into their consciousness but also is a necessary and vital form of grieving over the imminent loss of a safe and known past and the invasion of an unknown and uncertain future.

As noted, it is the letter of the law developers want strictly enforced. But, the letter of the law allows no sensitivity to either local conditions or extenuating circumstances, both of which may have consequences that are critical to the long-term environmental sustainability of a community. So, long after a developer is gone, the community is left to deal with the impacts of development, some of which strain at the fabric of place.

With this in mind, consider that coercive power is by belief—not force of arms or even capital. When people cease to accept their old beliefs, an empire can collapse almost overnight. However, "those who possess power will surrender it only when they see that the costs—physical or psychological—of retaining it are higher than the costs of losing it."[26]

When, however, coercive, political power eliminates public debate, it steals people's legal right to express how they feel, which causes their trust and emotional health to wither like a leaf in a hot wind. To be healthy, people not only must be allowed to grieve but also must be given permission to grieve for their perceived losses—a vital purpose of public debate.

Initially, however, we, as community members, often resist change because we are committed to protecting our existing values, representing as they do the safety of past knowledge in which there are no unwelcome surprises. We try to take our safe past and project it into an unknown future by skipping the present, which represents change and holds uncertainty, danger, and grief. But, mourning for ecological losses or even for a small, private, spiritual sanctuary, say a wooded hill near one's home, is a path that is neither simple nor predictable.

It is no surprise, therefore, that both our internal and external worlds often make grieving for the loss of our environment and our attachments to it a most difficult and uncertain process, in which the need to defend personal values and the feelings they engender is against the economics of development. Nevertheless, environmental and social losses are intermittent, chronic, cumulative, and without obvious beginnings and endings.

It is therefore necessary to encourage, support, and develop—not curtail—safe customs of grieving for environmental and social losses, those that alter the context of our lives just as surely as the loss of a person or pet. One of the best ways to encourage and develop a safe venue in which to grieve for our environmental and social losses (so that we may plan wisely for sustainable community development in the present for the present and the future) is to improve support for and the quality of existing public debate.[27] The kind of support necessary to understand and allow community members to deal with their grief emanating from the implementation of land use plans depends on an understanding of the human need for leisure and the time for reflection and processing of grief it provides.

Of Leisure and Community

The Chinese character for *leisure* is composed of two elements, which by themselves mean open space and sunshine. Hence, an attitude of leisure creates an opening that lets in the sunshine. Conversely, the Chinese character for *busy* is also composed of two elements, which by themselves mean heart and killing.

We tend to think of leisure, according to Brother David Steindl-Rast (a Benedictine monk), as the privilege of the well-to-do people. "But leisure," says Brother Steindl-Rast, "is a virtue, not a luxury. Leisure is the virtue of those who take their time in order to give to each task as much time as it deserves. … Giving and taking, play and work, meaning and purpose are perfectly balanced in leisure. We learn to live fully in the measure in which we learn to live leisurely."[28]

And, what does leisure have to do with sustainable community? To create a viable community and a shared vision for its sustainable future, people must give both their community and their vision for its future the time necessary for reflection, which means slowing down to recapture the *quality* of human relationships—the social glue of community. And, wisdom requires reflection.

The Slow Movement, which aims to address the issue of "time poverty" through making connections, recognizes this and, like the Living Building Challenge discussed previously, represents a host of values that are just now gaining traction. The Slow Movement has spawned many Slow Cities and created the emergence of places with Slow Food, Slow Money, and even Slow Schools.[29]

Yet, "As the tempo of modern life has continued to accelerate," says Jeremy Rifkin in his 1987 book *Time Wars*, "we have come to feel increasingly out of touch with the biological [and spiritual] rhythms of the planet, unable to experience a close connection with the natural environment."[30] Our human perception of time is no longer joined to the ebb and flow of the tides, the rising and setting of the sun, or the eternal parade of the seasons. We have instead created an environment governed by artificial time punctuated by electronic impulses from the heart of technology.

But, technology, contrary to the thinking of many people, is not the culprit in our ever-faster pace of life; the culprit is economics, where, as we have already said, time is money. Our social-environmental crises are based on the collision between scales of time—the fast scale of human modernity crashing into the slow scale of nature and Earth. The speeding internal clock of our constantly go-faster society is not only difficult to escape but also precludes most of us from thinking in terms of consciously varying the pace of our lives to find therein the hidden beauty, because our culture at large deems speed to be "productive." It is, nevertheless, imperative to break the sense of time as taskmaster because raising children, making close friends, and creating works of art all require various scales of time since they are practices in the aesthetics of relationship.

The aesthetics of relationship in scales of time is illustrated in a story Danny Hillis tells about the replacement of the gigantic oak beams in the ceiling of one of the dining halls at Oxford University. University officials were concerned that they would not be able to find lumber large enough and strong enough to replace the worn-out beams. But, ultimately the replacement beams were of the same quality as when the hall had been built 500 years earlier. How could this be?

"Simple," explained the university's forester. "When the dining hall was originally constructed 500 years ago, our predecessors were thoughtful enough, considerate enough, and farsighted enough to plant a grove of oak trees so that the university could, when necessary, replace the beams with others of the same quality." By planting the oaks all those centuries ago, one could say that the people who planted them not only understood and appreciated time in its various scales but also may well have seen time as a mystery to be contemplated, rather than a foe to be vanquished.[31]

Here it is important to understand that going slowly is not in opposition to speed, but rather is the middle path between *fast* and *inert*. Therefore, if a person will give up always looking beyond the task at hand, such as crafting a shared vision of a sustainable future, he or she inevitably finds that he or she not only has the required time for the task at hand but also finds that the task is done well the first time and thus actually saves time and money somewhere in the future.

Because true community and its shared vision are founded on the rare beauty of quality relationships, such a community and its shared vision for a sustainable future are possible only when we, as individuals, take the time to formulate our land use plans with the appropriate leisure to adequately fulfill our adult responsibilities as the social-environmental trustees of all generations. Such land use planning and its implementation are the cardinal responsibility of psychologically mature adults to *all* generations—a responsibility that requires an understanding of nature's inviolable biophysical principles.

Summary

Chapter 2 addresses the human components that make a sustainable community, such as a sense of place, history, and trust—which protects the members' right to grieve for the loss of cherished places due to the implementation land use planning. These elements are considered in terms of today's changing society, an understanding that is necessary to set the stage for Chapter 3, which is a discussion of the inviolable principles of biophysical sustainability.

Endnotes

1. Robert D. Kaplan. "Fort Leavenworth and the Eclipse of Nationhood." *The Atlantic Monthly* 278 (1996):75–78, 80–82, 85–85, 88–90.
2. Ibid.
3. Georg Feuerstein, Subhash Kak, and David Frawley. "The Vedas and Perennial Wisdom." *The Quest* 8 (1995):32–39, 80–81.
4. The foregoing discussion of architecture and history is based on James Howard Kunstler. "Home from Nowhere." *The Atlantic Monthly* 278 (1996):43–66.
5. Jonathan Hiskes. "The Case for Super-Ambitious Living Buildings. A Talk with Jason McLennan." 2010. http://grist.org/article/2010-09-30-jason-mclennan-living-buildings-interview/ (accessed December 8, 2012).
6. Ibid.
7. James Swan and Roberta Swan. *Dialogues with the Living Earth*. Quest Books, Wheaton, IL, 1996.
8. Matthew C. Marley and Alison K. Fields. "Seasonality of Moves and the Duration and Tenure of Residence: 2004." Issued July 2010. http://www.census.gov/prod/2010pubs/p70-122.pdf (accessed November 1, 2012).
9. Sarah van Gelder. "Beyond Greed and Scarcity." *YES! A Journal of Positive Futures* Spring (1997):34–39.
10. Caroline M. Myss. *Anatomy of the Spirit: The Seven Stages of Power and Healing*. Three Rivers Press, New York, 1996.
11. David C. Korten. "Money versus Wealth." *YES! A Journal of Positive Futures*, 2 (1997)
12. Susanna Kim. "Poll: Americans Need $150K Income to Feel Rich." 2011. http://abcnews.go.com/blogs/business/2011/12/poll-americans-need-150k-income-to-feel-rich/ (accessed June 13, 2012).
13. The foregoing discussion of money is based in part on (1) Korten, "Money versus Wealth"; and (2) Russ Beaton and Chris Maser. *Economics and Ecology: United for a Sustainable World*. CRC Press, Boca Raton, FL, 2012.
14. Henry Ford. http://thinkexist.com/quotation/the_highest_use_of_capital_is_not_to_make_more/11020.html (accessed June 13, 2012).
15. The discussion on the health of living in a community is based on the article by Mark Harris. "The Ties that Bind Are the Ties that Heal." *Vegetarian Times* August (1997):63–67.
16. Ibid.
17. Korten, "Money versus Wealth."
18. Wendell Berry. *Sex, Economy, Freedom, and Community*. Pantheon Books, New York, 1993.
19. Ibid.

20. Marc Luyckx. "The Re-enchantment of Politics." *YES! A Journal of Positive Futures* Winter (1997):16–17.
21. Ralph Waldo Emerson. http://www.quoteworld.org/quotes/4507 (accessed June 18, 2012).
22. Ronald L. Warren. *The Community in America* (2nd ed.). Rand McNally College, Chicago, 1972.
23. Margaret Mead. http://en.wikiquote.org/wiki/Margaret_Mead (accessed June 18, 2012).
24. "Abba Eban." *The Times* (London) (December 17, 1970).
25. William M. Adams, Dan Brockington, Jane Dyson, and Bhaskar Vira. "Managing Tragedies: Understanding Conflict over Common Pool Resources." *Science* 302 (2003):1915–1916.
26. George Monbiot. "Why We Conform." *Resurgence* 221 (2003):16–17.
27. The foregoing discourse is based on Chris Maser. "Grieving for Our Environmental/Social Losses." *Trumpeter* 13(4) (1996). http://trumpeter.athabascau.ca/index.php/trumpet/article/view/239/343 (accessed September 22, 2012).
28. Brother David Steindl-Rast. *Gratefulness and the Heart of Prayer: An Approach to Life in Fullness*. Paulist Press, Ramsey, NJ, 1984.
29. "The Slow Movement: Making a Connection." n.d. http://www.slowmovement.com/ (accessed November 20, 2012).
30. Jeremy Rifkin. *Time Wars: The Primary Conflict in Human History*. Simon and Schuster, New York, 1987.
31. (1) Jay Walljasper. "The Speed Trap." *Utne Reader*, March–April (1997):41–47; and (2) "Oak Beams, New College Oxford." n.d. http://atlasobscura.com/place/oak-beams-new-college-oxford (accessed June 18, 2012).

3

Nature's Inviolable Biophysical Principles and Land Use Planning

> We shall not cease from exploration, and the end of all our exploring will be to arrive where we started and know the place for the first time.
>
> **—T. S. Eliot**[1]

> Nothing in life is to be feared. It is only to be understood.
>
> **—Marie Curie**[2]

Change often occurs on the brink of disaster between need and fear—the need to know and the fear of knowing. On the one hand, we know we need to do things differently. On the other hand, we are terrified of facing the unknown, unfamiliar, uncomfortable unknown with all its impending uncertainties.

To change our direction for the future, however, we must suspend our conventional notions about change and nurture our ability to move from the level of the problem (where reaction, fear, and anxiety typically exist) to the level of the solution in which our beliefs, perceptions, feelings, assumptions, and expectations can be questioned and analyzed. With this expanded awareness, we can begin to create solutions that are free from these limitations. The problems we face are a matter of who we are consciously, and many people prefer to err repeatedly rather than let go of some cherished belief or notion, deified assumption, or staunchly defended position. Others err because they are habitually pessimistic in their outlook and thus unable to see alternative ways of viewing a problem.

Social-environmental sustainability demands that decision makers go beyond the immediate human valuation of a given natural resource to examine and disclose the fundamental issue of how its use will affect the long-term, biophysical sustainability of the ecosystem of which it is a component. One must also recognize and disclose the long-term, social-environmental issues that need to be addressed concerning the method by which a resource is extracted or modified. This is critical because the overall integrity of an ecosystem, its productive capacity, and the sustainability of its resources will determine the array of options passed forward to future generations.

To advance social-environmental integrity, all participating in land use planning must understand the inviolable biophysical principles set forth in this chapter, as well as their essential role in wise decision making. These principles form the underpinnings through which nature operates and the social limitations we must understand and accept if we are to participate with nature in a sustainable fashion. Moreover, social-environmental justice dictates that leaders be held accountable for the outcome of their decisions (while in a position of authority, rather than after the fact) because the effects of their decisions become the consequences for *all* generations—those present and those yet to come. This is particularly poignant in the face of an exploding human population, rapidly degrading ecosystems, and thus dwindling per capita resources.

Although these principles are presented in what seemed logical order, it is impossible to be definitive because each principle is an ever-interactive strand in the multidimensional web of energy interchange that constitutes our planet, its atmosphere, hydrolithosphere, and biosphere. Moreover, a different order can be found each time the principles are read, and each arrangement seems logical. And, because each principle affects all principles (like a water bed), every arrangement is equally correct in its overall function.

Principle 1: Everything Is a Relationship

The universe is a single, all-inclusive relationship constituted of an ever-expanding web of biophysical feedback loops, each of which is novel and self-reinforcing. Each feedback loop is a conduit whereby energy is moved from one place, one dimension, and one scale to another. All we humans do—ever—is practice relationships with the flow of energy within this web because the existence of everything is an expression of its relationship to everything else within the web through the continual exchange of energy. Beyond that, all relationships are forever dynamic, from the wear on your toothbrush from daily use to the rotting lettuce in the refrigerator. Herein is one of the foremost paradoxes of life: The ongoing process of change is a universal constant over which, much to our dismay, we have no control.

We, in our fear of uncertainty, are continually trying to hold the circumstances of our life in the arena of constancy through the decisions we make, as depicted in a snapshot—hence, the frequently used term *preservation* in regard to this or that ecosystem, this or that building. Yet, jams and jellies are more correctly termed *preserves* because they are heated during their preparation to kill all living organisms and thereby, it is hoped, preventing noticeable change (deterioration) in their consistency. Nevertheless, all systems are ultimately open because they are all part of a single cosmic relationship with energy, which tells us that a totally self-contained, closed-loop system—whether natural or artificial—is a physical impossibility.

Application of Principle to Land Use Planning

Although we know that everything is fundamentally in relationship to everything else, these relationships, as well as their exact nature, are frequently overlooked in the process of land use planning. Thus, understanding the context of both the biophysical and built environment is important when considering an alteration to the landscape, which must be at the scale of a single proposed development or a community-wide land use plan. In studying and considering the surrounding built environment, potential biophysically functional relationships can be discovered and, when recognized and understood, reflected in the design and function of the new development. In so doing, we follow and honor the principle of *everything is in relationship.*

Checklists are one way this process of comprehensive consideration is accomplished.[3] They help us remember, as well as manage, the multitude of variables that come into play when deciding to alter our cultural environment, which de facto alters our natural environment. For this approach to engender a sustainable design, it must be incorporated into a project from the earliest stages. Sustainable design is intended to produce buildings and landscapes that bring together a high positive regard for quality of life with the least ecological impact. Like all applications of the principles of sustainability, sustainable design aims at balancing economic, social, and environmental factors. In addition, the principles of sustainable design affect all phases of project development, from design, construction, operations, and maintenance to demolition and disposal.

Example 3.1: Industrial Symbiosis, Kalundborg, Denmark

Frequently cited as an exquisite example of sustainable development, the industrial symbiosis found in Kalundborg, Denmark, builds on the notion of open systems and associated interconnections and relationships. It is a good example to keep in mind when planning for any land use for which, throughout the process, we are seeking to make and discover a multitude of connections.

In Kalundborg, a series of industrial relationships has evolved over a period of 20 years. Beginning in 1970, it developed into a web of materials and energy exchanges among companies and the community. These relationships were originally prompted by a desire to save on costs by identifying income-producing uses for waste products. In time, the associated environmental benefits became obvious.

The Asnaes Power Station, Denmark's largest power station (coal fired), is at the center of the industrial symbiosis "hub" and is in relationship with Denmark's largest oil refinery, Statoil, the Novo Nordisk pharmaceutical plant, and the Gyproc plasterboard factory. "Over the last two decades, these partners spontaneously developed a series of *bilateral exchanges,* which also include a number of other companies. There was no initial planning of the overall network; it just evolved as a collection of one-to-one deals that made economic sense for the pairs of participants in each."[4] These relationships started to develop when Gyproc located its facility in Kalundborg to take advantage of the fuel gas available from Statoil.

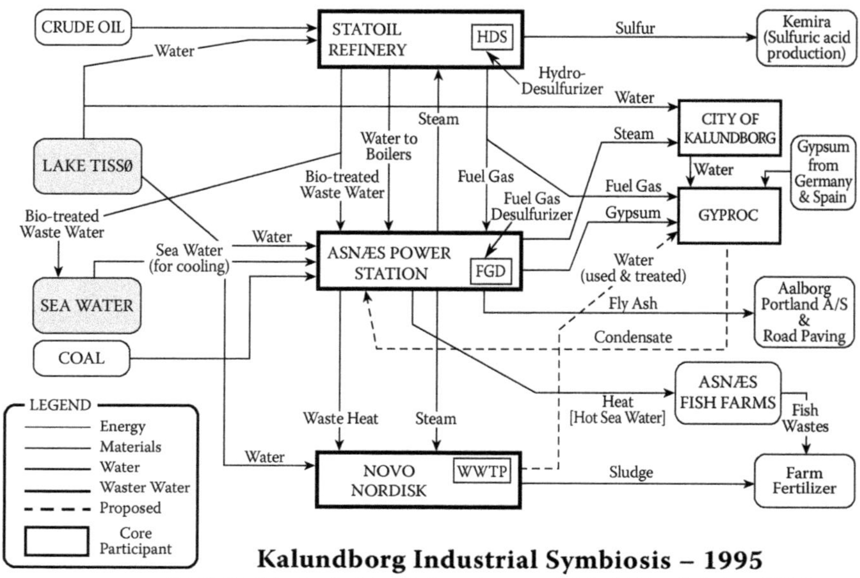

Kalundborg Industrial Symbiosis – 1995

Drawn by D. B. Holmes based on information from various sources, including L.K. Evans, N. Gertler, and V. Christensen

ENERGY FLOWS

Most of the energy from the Asnaes Power Station, as well as the Statoil refinery, as in many industrial operations, exits via the stack. In recognition of this waste, the two entities hatched a plan by which they could sell their "waste" to other entities. This occurred during the 1970s:

- Statoil agreed to provide excess gas to Gyproc.
- Excess steam from Asnaes was sent to the city for its new district heating system. The arrangement with the city resulted in the replacement of 3,500 oil furnaces (a significant nonpoint source of air pollution). Novo Nordisk and Statoil also became customers for this steam.
- Asnaes uses saltwater from the fjord for some of its cooling needs. The resulting by-product is hot saltwater, a small portion of which is supplied to a nearby fish farm's 57 ponds.[5]

MATERIALS FLOWS

As the mutually benefiting relationships grew, Novo Nordisk started the pattern of materials flows in 1976, matching the evolving energy flows at Kalundborg.

- Sludge from Novo Nordisk's processes and from the fish farm's water treatment plant is used as fertilizer on a nearby farm. This material exchange totals over 1 million tons (907,185 metric tons) per year.
- Desulfurized fly ash from Asnaes is used by a nearby cement company.
- Asnaes reacts the SO_2 in its stack gas with calcium carbonate, thereby making calcium sulfate (gypsum), which it sells to Gyproc, supplying two-thirds of the latter's needs.
- Statoil refinery's desulfurization operation yields pure liquid sulfur, which is trucked to Kemira, a sulfuric acid producer.
- Surplus yeast produced by Novo Nordisk's insulin production process is sent to farmers as pig food.[6]

This web of recycling and reuse has generated new revenues and cost savings for the companies involved and reduced pollution to air, water, and land in the region. In ecological terms, Kalundborg exhibits the characteristics of a simple food web: Organisms consume one another's waste materials, gaining energy as they become interdependent with one another.

This pattern of intercompany reuse and recycling has reduced air, water, and ground pollution; conserved water and other resources; and generated new revenue streams from the by-products exchanged.[7]

The elegance of this model in terms of reuse and recycling produced cost savings and revenues, as do other examples of sustainability. Through 1993, cost savings ranged around $120 million after taking into consideration the $60 million investment in infrastructure (to transport energy and materials).[8]

Principle 2: All Relationships Are Inclusive and Productive of an Outcome

It is often said that a particular piece of land is "unproductive" and needs to be "brought under management," and land use decisions are made accordingly. Here, it must be a clear that every relationship produces a cause that has an effect, and the effect, which is the cause of another effect, is the product. Therefore, the notion of an unproductive parcel of ground or an unproductive political meeting is an illustration of the narrowness of human valuation because such judgment is viewed strictly within the extrinsic realm of personal values, usually economics—not the intrinsic realm of nature's dynamics. Likewise, the conventional use of the term *waste* implies that something is of little to no value, when in fact all so-called waste ultimately becomes fuel for components of biophysical systems.

These narrow valuations not only illuminate how little we understand nature's inviolable dynamics but also should lead us to question the validity of our knowledge and our ability to adequately assess the economics associated with land use decisions.[9] After all, humans are not so powerful a natural force that we can completely destroy an ecosystem because it will continue, no matter what, to obey the biophysical principles that determine how it functions at a given point in time. Nevertheless, we can, and have, so severely altered an ecosystem that it is incapable of providing—for many decades and even for all time—those goods and services we require to sustain human life and society.

For example, the total surface area covered in paved roads in the United States precludes the soil's ability to capture and store water. In addition, the ocean's ability to sequester carbon dioxide (one of the main greenhouse gases) is being severely impaired due to the dramatically disrupted population dynamics of marine fishes by systematically overfishing too many of the top predators.[10] All of the relationships we affect produce an outcome or product. Whether the product is beneficial to humans or even amenable to our long-term existence is another issue.

Application of Principle to Land Use Planning

In recognition of this principle, which implies that all actions produce an outcome, the potential impact or outcomes of proposed land use actions therefore must be taken into account. This may take more time than some feel is acceptable, but like consensus decision making, the additional time taken will generally lead to solutions that are more durable. Consensus decision making, sometimes referred to "grassroots democracy,"[11] is a slower process than majority rule decision making but, by its very nature, allows more time for the reflection required when considering the consequences of our decisions.

Previously, we referenced a checklist approach. This is useful, but it represents our limitations in fully grappling with and comprehending the multitude of complex relationships inherent in any proposed land use action. If we acknowledge the potency and irreversibility of change, then perhaps a paradigm change at the level of society is needed first. If time for reflection played a more prominent role in the public process related to land use decisions, for example, we might be more cautious and respectful with every action we take in modifying nature by focusing more clearly on the greater good for *all* generations.

Example 3.2: The Slow Movement

Principle 2 represents a high-level structural understanding of nature. As such, how might this principle apply if we consider society and its institutions, including the economic system, education, food, land use, and so on? In answering this question, we must look at the new social paradigms or worldviews that honor this principle.

The Slow Movement was founded on the belief that people want greater connection with one another, their place, their food, transportation, and other social institutions, and that much of what ails society stems from broken connections. "The Slow Movement is a worldwide attempt to recapture this state of connectedness. The movement is gaining momentum, as more and more people recognize their discomfort with the fast pace and disconnected nature of their lives."[12]

It was not that long ago, for example, that people had a much greater connection to their food. Backyard gardens were common. Many had farm animals as part of their food supply. They were intimate with their source of food, and mealtimes were often social events.

An outgrowth of the Slow Movement is the Slow City Movement (Cittaslow International). Cittaslow certifies cities that meet certain requirements. Its manifesto states that it is looking for "towns where [people] are still curious of the old times, towns rich of theatres, squares, cafes, workshops, restaurants and spiritual places, towns with untouched landscapes and charming craftsman where people are still able to recognize the slow course of the Seasons and their genuine products respecting tastes, health and spontaneous customs."[13] The implications for land use planning at the community level are obvious. They pertain to the protection of scale, relatedness, promotion of the local economy, and environmental protection, such as open spaces and natural landmarks.

Slow Food is another outgrowth of the Slow Movement and since 1999 has now expanded to 100 towns and 10 countries worldwide. Slow Food "is an idea, a way of living and a way of eating. It is part of a global, grassroots movement with thousands of members in over 150 countries, which links the pleasure of food with a commitment to community and the environment."[14]

Again, the implications for land use are inherent in the idea of local food and protection of land for agricultural purposes, particularly organic agricultural use, which includes the idea of community gardens.

Principle 3: The Only True Investment in Our Global Ecosystem Is Energy from Sunlight

The *only true investment* in the global ecosystem is energy from solar radiation (materialized sunlight). Everything else is merely the recycling of already-existing energy. In a business sense, for example, one makes money (economic capital) and then takes a percentage of those earnings and *recycles* (= reinvests) them by putting them back into the infrastructure of the enterprise for maintenance of buildings and equipment. This reinvestment facilitates making a profit by protecting the integrity of the initial outlay of capital over time. In a business, one recycles economic capital *after* the profits have been earned.

Biological capital, on the other hand, must be "recycled" *before* the profits are earned to stay even. This means people need to forego some potential monetary gain by leaving enough of the ecosystem intact for it to function in a sustainable manner. In a forest, for instance, one leaves some proportion of the merchantable live, dying, and dead trees. The live trees gather solar radiation and produce living tissue that will subsequently die, and, together with the already-dead trees, rot, recycle into the soil, and thereby replenish the fabric of the living system. In rangelands, one leaves the forage plants in a viable condition so they can capture solar radiation to germinate, grow, mature, seed, die, decompose, and protect the soil from erosion as well as add organic material to the soil's long-term ecological integrity and productive capacity.

In contrast, people speak incorrectly about fertilization as an *investment* in a forest, grassland, or agricultural crop when in fact it is merely recycling chemical compounds that already exist on Earth. In reality, people are simply taking energy (in the form of preexisting chemical compounds) from one place and putting them in another for a specific purpose. The so-called investments in the stock market are a similar shuffling of preexisting energy.

When people "invest" money in the stock market, they are really recycling energy from nature's products and services acquired through human labor. The value of the labor is transferred symbolically to a dollar amount, thereby representing a predetermined amount of labor. Let us say, for instance, a person works for $10 an hour; then, a $100 bill would equal 10 hours of labor. Where is the investment? There isn't any investment, but there is a recycling of the energy put forth by one's labor and a symbolic recycling of the energy in the denomination of currency we spend.

Although one might argue that people invested their labor in earning the money, the fact is that whatever energy they put forth was merely a recycling of the energy they took in through the food they ate. Nevertheless, the energy embodied in the food may actually have simultaneously been a true investment and a recycling of already-existing energy.

It has long been understood that green plants use chlorophyll molecules to absorb sunlight and use its energy to synthesize carbohydrates from carbon dioxide and water. This process is known as photosynthesis, where *photo* means "light," *synthesis* means the "fusion of energy," and the combination is the basis for sustaining the life processes of all plants. The energy is derived from the sun (an original input) and combined with carbon dioxide and water, as well as nutrients from the soil in which the plants grow (all existing chemical compounds), to create a renewable source of usable energy. This process is analogous to an array of organic solar panels—the green plant.

Put another way, the plant (an array of solar panels) uses the green chlorophyll molecule (a *photoreceptor,* meaning receiver of light) to collect light from the sun within chloroplasts (small, enclosed structures in the plant that are analogous to individual solar panels). Then, through the process of photosynthesis, the sun's light is used to convert carbon dioxide and water to carbohydrates as usable energy for the plant (which is partially sustained by chemical nutrients in the soil, where, incidentally, most of the water is also derived), a process that is comparable to converting the sun's light in solar panels (derived from recycled energy in the form of materials) on the roof of a building into electrical energy (electricity) for our use. These carbohydrates in turn are partly converted energy from the sun—a new input of energy into the global ecosystem—and partly the recycling and reuse of existing energy from the amalgam of carbon dioxide and water, a process sustained in part by chemical nutrients from the soil.[15]

When, therefore, we eat green plants, the carbohydrates are converted through our bodily functions into different kinds of energy. By that is meant the energy embodied in green plants—which is partly a true investment of energy—is altered through digestion into the various types of energy our bodies require for their physiological functions. The *excess* energy (that not required for physiological functions) is expelled as bodily waste or expended in the form of physical motion, such as work. On the other hand, it is different when eating meat because the animal has already used the sun's contribution to the energy matrix in its own bodily functions and its own physical acts of living, so all we get from eating flesh is recycled energy.

Application to Land Use Planning

The primacy of solar energy in Earth's life support system can and must be accounted for in land use planning practices and regulations. One way this can be done is through the regulation of solar access. Solar access, in turn, is divided into two distinct areas: solar easements and solar rights.

Where "solar easements" are protected, the ability of one person's property to receive sunlight across property lines without obstruction from another's property, such as buildings, foliage, or other impediments, is enabled. On the other hand, local municipalities that protect "solar rights" allow for the installation of solar energy systems on residential and commercial property, subject to such private restrictions as covenants, conditions, restrictions, bylaws, and condominium declarations.

Nevertheless, neither sunlight nor solar energy is considered common property (see previous discussion in Chapter 1 of the Seventh Generation Amendment to the U.S. Constitution). "The United States has held that there is no common law right to sunlight. This requires that specific statutory authority be established to protect the rights of solar users in terms of their ability to install a solar energy system on their property, and after that system is installed, to protect their access to sunlight so the system remains operational."[16]

As we know, local government has the authority to adopt solar access policies within the context of a local, comprehensive plan, which typically includes how land is to be used. Regulations emerging from the adopted land use plan can then include such things as provisions that would enable developers to provide southern exposures to as many buildings and lots as possible and thereby maximize their access to sunlight. In addition, landscaping, including trees and other major vegetation, as well as buildings, would need to be placed in a manner that minimizes the shadows cast on human-occupied structures.

This type of regulation would also enable a developer to seek an unobstructed solar sky above neighboring properties through solar easements, which are defined as "restrictions on adjoining lots that would prohibit intrusions into the solar sky space, such as another building or trees. A restrictive covenant can accomplish this as well by requiring that no solar energy collector … be shaded by any building, vegetation, or obstruction between certain hours on a certain date of any year."[17] Similarly, landscape ordinances can be designed to promote the type of vegetation that enhances access to solar energy rather than limiting it.

Example 3.3: Municipal Ordinances

Many cities in the United States have adopted land use regulations that provide for solar access. With growth in the solar energy industry, more and more land use regulatory agencies are following suit. The following cities have adopted such land use ordinances, which have been collected by the American Planning Association and contained in the APA Policy Guide:

- Ashland (Oregon), City of. 2011. *Municipal Code.* Title 18, Land Use; Section 18.70, Solar Access.
- Boulder (Colorado), City of. 2011. *Boulder Revised Code.* Title 9, Land Use Regulation; Chapter 9–9, Development Standards; Section 9-9-17, Solar Access. Fort Collins, Colo.: Colorado Code Publishing Company.
- Clackamas (Oregon), County of. 2011. *County Code.* Title 12, Zoning and Development Ordinance; Section 1000, Development Standards; Part 1018, Solar Balance Point/Infill Ordinance, and Part 1019, Solar Access Permit Ordinance.
- Fort Collins (Colorado), City of. 2011. *Land Use Code.* Article 2, Administration; Division 2.8, Modification of Standards; Section 2.8.2, Modification Review Procedures. Article 3, General Development Standards; Division 3.2, Site Planning and Design Standards; Section 3.2.3, Solar Access, Orientation, Shading. Article 5, Terms and Definitions; Division 5.1, Definitions. Fort Collins, Colo.: Colorado Code Publishing Company.
- Laramie (Wyoming), City of. 2009. *Code of Ordinances.* Title 5, Business Taxes, Licenses and Regulations; Chapter 5.58, Solar Rights Permit System. Tallahassee, Fla.: Municipal Code Corporation.
- Prairie du Sac (Wisconsin), Village of. 2008. *Code of Ordinances.* Title 10, Land Use Regulations; Chapter 8. Solar Access.

- San Luis Obispo (California), City of. 2011. *Municipal Code*. Title 16, Subdivisions; Chapter 16.18, General Subdivision Design Standards; Section 16.18.170, Easements for Solar Access. Seattle, Wash.: Code Publishing Company, Inc.
- Santa Barbara (California), City of. N.D. *Solar Access Packet*. Municipal Code Chapter28.11, Protection and Enhancement of Solar Access.[18]

Principle 4: All Systems Are Defined by Their Function

The behavior of a system—any system—depends on how its individual parts interact as functional components of the whole, not on what an isolated part is doing. The whole, in turn, can only be understood through the relationships, the interaction of its parts. The only way anything can exist is encompassed in its interdependent relationship to everything else, which means an isolated fragment or an independent variable can exist only on paper as a figment of the human imagination—of which it is, nevertheless, an interactive part.

In reality, all systems are interdependent and thus rely on their pieces to act in concert as a functioning whole. This being the case, no individual piece can stand on its own *and* simultaneously be part of an interactive system. Thus, *there neither is nor can there be an independent variable* in any system, be it biophysical, mechanical, or even intellectual, because every system is interactive by its very definition as a system.

What is more, every relationship is constantly adjusting itself to fit precisely into other relationships that, in turn, are consequently adjusting themselves to fit precisely into all relationships, a dynamic that *precludes* the existence of an independent variable, which also means that *nothing can be held as a constant value beyond the number one* (the universal common denominator) because to do so would necessitate the detachment of the thing in question from the system as an independent variable. Therefore, all relationships are constituted by additions of *one* in all its myriad forms, from quarks, atoms, molecules, and proteins, which comprise the building blocks of life, to the living organisms, which collectively form the species and biotic communities. The only way the number one can exist, as the sole representative of any form on Earth, is to be the last living individual of a species—something intimated on the tribal level by James Fennimore Cooper's 1826 book, *The Last of the Mohicans*[19]—because extinction is forever.

Therefore, to understand a system as a functional whole, we need to understand how it fits into the larger system of which it is a part and so gives us a view of systems supporting systems supporting systems supporting systems ad infinitum.

Application of Principle to Land Use Planning

We recognize this principle in both the physical and cultural context of land use planning when we carefully study a contemplated or planned modification to the landscape. This kind of examination helps us to identify systems within systems and associated interrelationships. While we know that change is constant, it is incumbent on us to do this sort of analysis routinely as part of responsible land use planning.

Example 3.4: Sustainable Seattle Community Idea and Project Proposal Checklist

One of the earliest checklists emerging from the nongovernmental sector was Sustainable Seattle's Sustainable Community Development checklist. This checklist and similar ones attempt to give assurances that context is examined and analyzed as part of a proposed land use action.

Another example is the Minnesota Pollution Control Agency, which adopted and then adapted the Sustainable Seattle checklist. It is divided into five areas: (1) community development, (2) ecological health, (3) economic health, (4) social equity and connections, and (5) trade-offs and the long term. In each category, questions are posed to which answers can range from strongly disagree to strongly disagree. These questions illuminate systems within systems and give value to dimensions frequently overlooked when considering a land use proposal.

For example, in the community development category, the checklist asks, under the general heading of "Inclusiveness," "Does the proposal encourage the participation of all affected people in the decision-making process?" And, under "Mobility": "Does the proposal for transportation and information access within and without the community foster alternatives to single-occupancy car use?" And, under "Community Identity": "Does the proposal help citizens feel a sense of belonging to the community and a sense of ownership in it?"

The "Ecological Health" section of the checklist looks at how well the proposal takes ecological opportunities and limitations into account. For example, under "Ecosystems": "Does the proposal improve the quality of watersheds, plant and animal life, with attention paid to wildlife corridors?" And, under "Energy": "Does the proposal promote a reduction in the use of energy, renewable energy, and greater efficiency?" Under "Peace and Quiet": "Does the proposal reduce noise pollution?" In the "Connections, Trade-offs, and Long-Term" section of the checklist, under "The Seven Generation Test," it asks: "Does the proposal consider its impacts on the community 175 years from now?" And, under "Trade-offs in the Community": "Does the proposal seek to meet social, economic, and environmental goals simultaneously and, where it cannot, does it make reasoned and balanced trade-offs, informed by the community's core values?"[20]

While completing such a checklist may not result in quantitative evidence for decision making, it clearly provides qualitative information, something that, in sustainable community development, is vital in the all decision making. This is not meant to preclude quantitative data as an input to the process, rather as an "enhancement" to the process.

Principle 5: All Relationships Result in a Transfer of Energy

Although technically a *conduit* is a hollow tube of some sort, the term *conduit* is used here to connote any system employed specifically for the transfer of energy from one place to another. Every living thing, from a virus to a bacterium, fungus, plant, insect, fish, amphibian, reptile, bird, mammal, and every cell in our body, is a conduit for the collection, absorption, transformation, storage, transfer, and expulsion of energy. In fact, the function of the entire biophysical system is tied up in the collection, absorption, transformation, storage, transfer, and expulsion of energy—one gigantic, energy-balancing act, or perhaps more correctly energy-juggling act.

For example, prior to the invasion of the "New World" by Europeans, human impact on the environment was not simply a process of continually increasing change in response to the growth of the indigenous populations. Instead, the landscape was given time to rest and recover as people moved about; cultures collapsed; populations declined because of periodic starvation, disease, and war; and habitations were abandoned.

Human activities may be constructive, benign, or destructive, all of which are subjective concepts based on human values, but change—the dance of energy—is continual, albeit at various rates and in various directions. For example, although plants have no ear with which to hear the increasing noise of human machinery, such as automobiles, airplanes, trains, oil and gas wells, and other anthropogenic-derived noise, animals do. As a result of the increasing noise pollution, biological communities are being restructured, which in turn alters the function of whole ecosystems.[21] Moreover, all changes are, in addition, cumulative. Even mild, slow change can show dramatic effects over the long term.

Application of Principle to Land Use Planning

This principle implies the importance of being aware of change over time, as it can be slow and unnoticeable but important to ecosystem health, cultural well-being, and thus the survival of all life. Therefore, identifying measurable indicators is an important inclusion in the process of land use planning for both the short and long term. Regular measurement of change, enabled by identified critical indicators, is one way of applying this principle to land use planning.

Indicators show us how well a system is working and help us recognize problems in a system. Indicators vary, depending on the type of system they monitor. They should be relevant, easy to understand, and reliable and provide information while there is still time to act.

Indicators of sustainability must address the issue of the community's carrying capacity; highlight links between social, economic, and environmental well-being; be usable by people in the community; and advance local sustainability, but not at the expense of other communities.

Like the checklists mentioned, indicators should align with key components of sustainable community: ecological integrity, quality of life, economic security, empowerment, and responsibility.[22]

Example 3.5: Sustainable Seattle Indicators Report

Sustainable Seattle's landmark 1998 regional indicators report stands as a model for other communities.[23] This guide contains 40 indicators of social, environmental, and economic health and was motivated by questions, such as the following: How do we balance concerns for social equity, ecological integrity, and economic vitality? And, how do we create a livable community today while ensuring a healthy and fulfilling legacy for our children's children?

Indicators identified fall into five categories: environment, population and resources, economy, youth and education, and health and community. Within each category are measurable indicators representing significant and, in some cases, worrisome trends. Under "environment," for example, we find salmon run, vegetative cover, soil erosion, air quality, pedestrian- and bicycle-friendly streets, open spaces, and impervious surfaces. These measurable elements enable us not only to plot trends but also to serve as a guide in evaluating development proposals.

Although all categories of indicators developed by Sustainable Seattle have relevance for land use, we have highlighted the Environment category only. Within the realm of sustainability, however, all factors falling under the specific categories of economy, population and resources, education, and community and health form interconnected relationships of differing scales.[24]

Principle 6: All Relationships Are Self-Reinforcing Feedback Loops

Everything is connected to everything else in a cosmic web of interactive feedback loops, all entrained in self-reinforcing relationships that continually create novel, never-ending stories of cause and effect. Everything, from a microbe to a galaxy, is defined by its ever-shifting relationship to every other component of the cosmos. Thus, "freedom" (perceived as the lack of constraints) is merely a continuum of fluid relativity. In contraposition, every relationship is the embodiment of interactive constraints to the flow of energy—the very dynamic that perpetuates the relativity of freedom and thus all relationships, which by definition precludes absolute freedom.

Hence, every change (no matter how minute or how grand) constitutes a systemic modification that produces novel outcomes. A feedback loop, in this sense, comprises a reciprocal relationship among countless bursts of energy moving through specific strands in the cosmic web that cause forever-new, compounding changes at either end of the strand, as well as every connecting strand.[25] And here, we humans often face a dichotomy with respect to our interests.

On the one hand, while all feedback loops are self-reinforcing, their effects in nature are neutral because nature is impartial with respect to consequences. We, on the other hand, have definite desires as far as outcomes are involved and thus assign a preconceived value to what we think of as the end result of nature's biophysical feedback loops—of which our thought, decisions, and actions are an integral part, relative, for example, to such things as clean air in a city versus polluted air.

Application of Principle to Land Use Planning

If, indeed, all sustainability practices, including land use planning for sustainable development, are about emulating nature, then creating and enhancing feedback loops in our process and outcomes is key. Thus, when planning at any scale, seeking to discover and understand existing feedback loops, building on them, or expanding on them has merit.

At the scale of community, the application of this principle can be seen in the eco-districts model, local living economies, ecomunicipalities, and Transition towns.

Example 3.6: Ecomunicipalities

An *ecomunicipality* is an area, not necessarily defined by geopolitical boundaries, that has adopted the values of ecological and social justice, as well as "The Natural Step." The development of ecomunicipalities originated in Sweden, where more than 70 municipal governments have accepted varying principles of sustainability in their operations, as well as community-wide decision-making processes.

The concept of an ecomunicipality differs from other notions of sustainable development (such as green building, sustainable agriculture, and alternative energy) in that the focus is on community involvement and social transformation, as well as the use of a whole-systems approach.[26] An ecomunicipality is one that recognizes that principles of sustainability must guide all decisions made by government and result in actions that model the way for other entities in the community, including businesses, nonprofit organizations, and so on.

The ecomunicipality-systems approach focuses on the relationship among the parts of community as well as the parts themselves. It assumes, for example, that long-term financial well-being is linked to long-term social capital or "the networks of relationships among persons, firms, and institutions in a society, together with associated norms of behavior, trust, cooperation, etc., that enable a society to function effectively."[27]

Sustaining long-term social capital means providing the "vessel" that nurtures such networks. Social capital represents active links among people, which includes mutual understanding, trust, shared values, and behaviors that forge relationships and make cooperation possible. How is social capital linked to financial capital? Building financial capital requires having talented people in the community, and one of the best ways to attract talented people is to have an economic development strategy that creates a healthy housing supply, excellent health care, access to education and culture, and a sense of community. Without these, individuals are less likely to take business-oriented risks. Putting it simply, if social capital is low or marginal, the rate at which financial capital can grow is diminished.[28]

"Capitalizing on assets" means looking at all so-called waste as a commodity. Waste is becoming a commodity because it can be converted, through new and emerging technologies, into fuels, energy, and products formerly reliant on extracted resources. This concept emulates nature since in nature all waste is fuel for other living systems. In northwestern Wisconsin, for example, Elkhorn Industries in Superior, Wisconsin, proposed an ecoindustrial park that includes the following: the conversion of wood scraps, everyday garbage, waste heat, sewage sludge, and grain dust from nearby Superior industries to new products. The park would include a gasification unit and power generation plant, along with manufacturing facilities, that would produce fiberboard, ethanol (from synthetic gas piped into a bioreactor that has bacteria in it that transforms the synthetic gas), electricity (from heat and from feeding some "syngas" into fuel cells), and tomatoes (from greenhouses heated with heat and energized with carbon dioxide from the bacteria).[29]

The design of ecoindustrial parks is based on the idea that any material produced by manufacturing that is not part of a product can be a feedstock for another manufacturing process.[30] "Capitalizing on existing assets" also means keeping our most precious assets—people, water, air, soils, and natural systems—healthy. As part of the systems thinking dimension of the ecomunicipality model, the functional conditions of systems represent the foundation of sustainability.

Thus, maintaining the "system," in this case natural, biophysical systems on which we depend for our lives and livelihoods, ultimately means being cognizant of and being accountable for the substances we introduce into the environment, which include materials from the earth's crust introduced into the parts of the universe inhabited by living organisms (biosphere). These include fossil fuels, metals, and other minerals. Why is this important? Once introduced, the effects are immediately out of our control. A few examples are waste from nuclear energy, carbon dioxide from burning fossil fuels that acidifies the oceans, heavy metals that poison the soil, and sulfur dioxide air pollution. Such care is critical because many of these substances have been shown to have harmful effects on living cells, including human, animal, and plant cells. Further, we cannot eliminate these substances, once released, such as we observe with carbon dioxide and its effect on climate. This idea represents the first "system condition" of The Natural Step.

The second "system condition" calls for the ultimate elimination of socially produced substances that do not break down into nontoxic components. *Nontoxic* means that plant or animal cells can metabolize it without disruption of normal functions. These substances also tend to disperse and accumulate in various tissues of living organisms. Further, many safe substitutes for these synthetics, including new nontoxic substances and alternative processes, are now being discovered. The harmful health effects of mercury contamination found in the waters and fish of many lakes and rivers in the United States, as well as the oceans worldwide, have demonstrated an example of the need to fulfill this system condition.

The third system condition calls for the ultimate elimination of our manipulation of natural systems, such as overharvesting and other major disturbances. These systems provide free services that are essential for our lives and livelihoods. Included in these services are clean water, clean air, fertile soils, productive oceans, pollination, climate regulation, solar energy, detoxification and recycling of human waste, and a supply of drinking water that is stored part of the year in clean snowpack. Further, according to Fritjof Capra, physicist and systems theorist, what is maintained in nature is "not individual trees or organisms, but a complex web of relationships [among] them" (p. 159).[31] Protecting this structure, or the web of relationships, among such things as society, economy, and environment is key to the concept of ecomunicipality and will help to sustain life on Earth.

The fourth system condition, without which the first three become meaningless, is the fair and efficient use of resources with respect to meeting needs worldwide. Without basic human needs being met, the kind and scale of change proposed by the first three system conditions become impossible. In other words, if a person's attention is dominated by simply finding food or adequate shelter, the person cannot participate in a community seeking to ensure social and economic stability tied to the first three system conditions.

Communities in North America and Europe, ranging in size from villages of 300 to cities of 700,000, have become ecomunicipalities (e.g., Stockholm). In Sweden, over 70 municipalities have officially become ecomunicipalities. They have formed a national association of ecomunicipalities to assist one another and work to influence national policy. Whistler, British Columbia, Canada, was awarded first place in an international competition for sustainable communities endorsed by the United Nations.

On the East Coast of the United States, Lawrence Township, New Jersey, a suburb of Trenton and New York City, has joined the community of ecomunicipalities, as has the New England city of Portsmouth, New Hampshire. These emerging ecomunicipalities and others, such as Vandergrift, Pennsylvania, have formed the North American Eco-Municipality Network and hold periodic conference calls to assist one another. Generally, citizen initiative drove these efforts to become "green" or sustainable communities.[32]

No longer interested in "quick fixes," these entities are taking a longer-term systems approach to staying in business. This systems approach is an inherent part of the ecomunicipality concept and foundational to the thoughts and precepts set forth in *The Necessary Revolution*, a book by Peter Senge, Joe Laur, and other.[33]

Principle 7: All Relationships Have One or More Trade-offs

All relationships have trade-offs that may not be readily apparent or immediately understood. Take the springtime ozone hole over Antarctica as illustrative. It is finally shrinking after years of growing. As the hole grew in size due to the human-induced, ozone-destroying chemicals in the stratosphere, the risk of skin cancer increased because more ultraviolet radiation reached Earth. Although today the good news is that the ozone hole is now shrinking and, through a complicated cascade of effects, could fully close within this century, what about tomorrow? Because the hole in the stratospheric ozone layer does not absorb much ultraviolet radiation, it keeps the temperature of Antarctica much cooler than normal. A completely recovered ozone layer, on the other hand, could significantly boost atmospheric warming over and around the icy continent and ostensibly augment its melting.[34] In this case, what is good for humans may not be good for Antarctica and vice versa.

Application of Principle to Land Use Planning

Land use planning always involves compromise and trade-offs, the knowledge of which is incorporated into the process by asking, What are the trade-offs this action implies? Such knowledge is important, as well as the analysis of the impacts of the trade-off—social, environmental, and economic. To be clear, in this context, we will use the term *trade-off* to refer to something lost, which people would rather not be lost, for something gained, which has obvious merit. An example that relates to land use is the introduction of disincentives to use one's car for commuting and incentives for using mass transit or bicycles. Lost are convenience and speed; gained are cleaner air, less reliance on petroleum, reduced need for new roads, and reduced public costs.

Example 3.7: Well-Considered Trade-offs in Salt Lake City, Utah

Salt Lake City, Utah, is just one example of cities in the United States that have a noteworthy sustainability initiative under way. Here are a few examples of land use decisions that involved trade-offs:

1. Salt Lake City protects lands that are designated urban forests.
   Costs: Like other public land holdings, this means fewer acres (hectares) available for private development, both commercial and residential. This also means fewer acres (hectares) accruing tax revenue.
   Benefits: Salt Lake City's forest has a direct impact on environmental quality, public safety, and city revenues. Currently, there are 91,221 trees along city streets and in parks. These trees lower the costs of air conditioning and heating by 30 to 40 percent. When shaded by trees, the irrigation requirements of turf can be lowered by 20 percent, helping conserve the state's precious water resource.[35]
2. The city has a multiprong effort aimed at mitigating the rate at which climate change accelerates. Such efforts have a direct impact on land use since the amount of land dedicated to transportation typically represents a significant portion of the whole.
   One such effort engages citizens with the state of Utah's Clear the Air Challenge/Drive Down Your Miles program.[36] This is a straightforward message to citizens to reduce the use of their automobiles and make fewer single-person automobile trips.
   Costs: There is less automobility, thus less convenience, and greater demand on planning to satisfy life's necessities.
   Benefits: Last year, Utah eliminated 3 million single-person vehicle trips and thus averted 2 million miles (3,218,688 kilometers) of travel and reduced emissions by 1.7 million pounds (771,107 kilograms). A related benefit is cleaner air and thus reduced health risk. In terms of land use, this initiative could ultimately reduce the need for lands dedicated to roads.
3. Another initiative pertains to community gardens and urban farming. This use of the land is becoming more important as communities recognize the value of local-source economies, such as locally grown food. The city has taken measures to increase local food production through community gardens and urban farming. Actions taken by the city include
   - The city works directly with local nonprofit organizations to develop gardens within city limits. Wasatch Community Gardens, for example, is a community-based nonprofit that has served the community for over 20 years with a mission to empower people of all ages and incomes to grow and eat healthy, organic, local-grown food.
   - The city now allows poultry to be raised on privately owned land.
   - The city has installed the People's Portable Garden.
   - The city assisted in the purchase of land and a conservation easement to protect the 4th E Community Garden.

   Benefits: There is a greater supply of locally grown, organic food and its associated personal health advantages and a greater chance for community members to come together in productive, meaningful work.

Costs: There is lost revenue to retail grocers and, consequently, lost opportunities for traditional "economic development" with respect to the Chamber of Commerce.[37]

There are exquisite examples of land use planning where the trade-offs were not recognized and thus enormous costs accrued as a result. One example pertains to the Army Corps of Engineer's practice of canalizing waterways to "flood-proof" urbanized areas along the Mississippi River. While floodwaters are subsequently contained locally, their velocity is intensified by such restrictions, and the increased probability of flooding is transported downstream. Moreover, the floodplain cannot perform its biophysical functions of slowing the water, collecting sediment, and acting as winter habitat for fish and other wildlife, which says nothing about the subsequent loss of wetlands for water storage and wildlife habitat.[38]

Principle 8: Change Is a Process of Eternal Becoming

Change, as a universal constant, is a continual process of inexorable novelty. Although it is a neutral dynamic of eternal novelty, our human valuation of change invariably renders a verdict of either/or: good as opposed to bad, creation as opposed to destruction. Change is a condition along a continuum that may reach a momentary pinnacle of harmony within our senses, but then the very process that created the harmony takes it away and replaces it with something else—always with something else. Change requires constancy as its foil to exist as a dynamic process of eternal becoming. Without constancy, change could neither exist nor be recognized.

Immediate Change

We all cause change of some kind every day. I (Chris) remember a rather dramatic one I made inadvertently along a small stream flowing across a beach on its way to the sea. The stream, having eroded its way into the sand, created a small undercut that could not be seen from the top. Something captured my attention in the middle of the stream, and I stepped on the overhang to take a better look, causing the bank to cave in and me to have a really close-up view of the water. As a consequence of my misstep, I had both altered the configuration of the bank and caused innumerable grains of sand to be washed back into the sea from whence they had come several years earlier riding the crest of a storm wave.

Understanding Historical Change

Whereas this example is a small, personally created change in an infinitesimal part of the world, others are of gigantic proportions in their effects. People of civilizations that collapsed centuries ago are a good example of such gargantuan effects because they were probably oblivious to the impact that could be wrought by long-term shifts in climate. Although not likely to end the debate regarding what caused the demise of the Roman and Byzantine Empires, new data suggest that a shift in climate may have been partly responsible. The plausibility of this notion has been given a scientific boost of credibility through studying the stalactites of Soreq Cave in Israel.[39]

Stalactites are the most familiar, bumpy, relatively icicle-shaped structures found hanging from the ceilings of limestone caves. They are formed when water accumulates minerals as it percolates through soil before seeping into a cave. If the water's journey takes it through limestone, it typically leaches calcium carbonate and carbon dioxide in its descent. The instant the water seeps from the ceiling of a cave, some of the dissolved carbon dioxide in the fluid escapes into the cave's air. This gentle, soda-pop-like fizzing process causes the droplet to become more acidic and so results in some of the calcium carbonate crystallizing on the cave's ceiling, thereby initiating a stalactite. As this process is performed repeatedly, the separation of calcium carbonate from within the thin film of fluid flowing down its surface allows the stalactite to grow. The procedure is so slow it typically takes a century to add 0.4 inch to a stalactite's growth.[40]

By using an ion microprobe, it has become possible to read the chemical deposition rings of the Soreq Cave stalactites with such precision that even seasonal increments of growth can be teased out of a given annual ring. The results indicate that a prolonged drought, beginning in the Levant region as far back as 200 years BCE and continued to AD 1100, coincides with the fall of both empires.[41] (Levant is the former name of that region of the eastern Mediterranean that encompasses modern-day Lebanon, Israel, and parts of Syria and Turkey.) Although determining why civilizations collapse is always more complicated than one might imagine, an inhospitable shift in climate might well be part of the equation that either forces people to adapt by changing their behavior or eliminates them—the challenge squarely before today's decision makers in all countries in all levels of government.

Change as a Historical Continuum

People began to have an influence on Earth's climate thousands of years before the Industrial Revolution and the burning of fossil fuels to power machinery commenced spewing carbon dioxide into the air. Clearing land through the use of fire, first for hunting and then for farming, removed trees that otherwise would have sequestered carbon dioxide rather than allowing it to escape into the atmosphere. As well, the growing human population and the farming practice of disturbing the soil also allowed sequestered carbon to escape. During these early millennia of human interaction with their environment, their activities are estimated to have released about 386 billion tons (350,173,310 metric tons) of carbon into the atmosphere by 1850. Between 1850 and 2000, however, the Industrial Revolution and the burgeoning number of people added 485 tons (440 metric tons) of carbon, primarily from burning fossil fuels, which more than equaled in a century and a half what had previously taken humankind eight millennia to do.[42]

The foregoing set the stage for the historical record of the Asian monsoon's activity, which is archived in an 1,800-year-old stalagmite found in Wanxiang Cave in the Gansu Province of north-central China. Mineral-rich waters dripping from the cave's ceiling onto its floor year after year formed the stalagmite (a mirror image of a stalactite) that grew continuously for 1,800 years, from AD 190 to 2003. Like trees and the stalactites in the Soreq Cave of Israel, stalagmites have annual growth rings that can provide clues about local environmental conditions for a particular year. Chapters in the Wanxiang Cave stalagmite, written over the centuries, tell of variations in climate that were similar to those of the Little Ice Age, Medieval Warm Period, and the Dark-Age Cold Period recorded in Europe. Here, the critical point is that the warmer years were associated with stronger East Asian monsoons. (A monsoon is a large-scale system of wind that blows seasonally in opposite directions and thus determines the climate of a large region, such as the summer period of heavy rainfall over Southeast Asia.)

By measuring the amount of oxygen-18 (a rare form of "heavy" oxygen) in the growth rings of the stalagmites, the years of weak summer monsoons with less rain can be pinpointed due to the large amounts of oxygen-18 in the rings. The information secreted within the life of the stalagmite tells the story of strong and weak monsoons, which in turn chronicle the rise and fall of several Chinese dynasties. This is an important deliberation because monsoon winds have for centuries carried heavily rain-laden clouds northward from the Indian Ocean every summer, thereby providing nearly 80 percent of the annual precipitation between May and September in some parts of China—precipitation critical to the irrigation of crops.

In periods when the monsoons were strong, dynasties, such as the Tang (618–907) and the Northern Song (960–1127), enjoyed increased yields of rice. In fact, the yield of rice during the first several decades of the Northern Song dynasty allowed the population to increase from 60 million to as many as 120 million. But, periods of weak monsoons ultimately spelled the demise of dynasties.

Data from the stalagmite indicate that the strength of past Asian monsoons was driven by the variability of natural influences—such as changes in solar cycles and global temperatures—until 1960, when anthropogenic activity appears to have superseded natural phenomena as the major driver of the monsoon seasons from the late twentieth century onward. In short, the Asian monsoon cycle has been disrupted by human-caused climate change[43]—based on decades of symptomatic decision making. Here, an observation by the British biologist Charles Darwin is apropos: "It is not the strongest of the species that survive, nor the most intelligent, but *the one most responsive to change*" [emphasis added].[44]

Application of Principle to Land Use Planning

When we recognize change as a constant, we must then remember to build flexibility and adaptability into all land use plans, especially in light of significant and sudden unexpected change such as natural disasters. Having established indicators to measure progress toward achieving established goals will aid in adaptation, when change requires a response.

Example 3.8: City of Santa Cruz, California, 1989

I (Jane) was a city planner for the City of Santa Cruz from 1981–1990. In 1989, a 7.1 earthquake was centered approximately 7 miles (11 kilometers) to the north of Santa Cruz, rocked the city. Several months prior to the earthquake, the City had adopted a 20-year Comprehensive Plan, which took nearly 5 years to complete. Therefore, immediate revisions to the adopted plan were necessitated by sudden, unexpected, and impactful change brought about by an earthquake.

A significant portion of the recently adopted plan focused on the downtown with its many historic masonry structures, most of which had not been properly retrofitted with rebar, as required by state law, to prevent damage from seismic activity. "In Santa Cruz, the earthquake pierced what many called the soul of the city—the downtown. In less than 15 seconds, the historic and eclectic Pacific Garden Mall was reduced to rubble. … Countywide, there was an estimated $1 billion in losses and 206 businesses were damaged or destroyed along Pacific Avenue."[45] Sixty buildings were damaged, and many needed to be demolished.

This was a great lesson in accepting the losses and moving forward. Businesses, in particular, needed the greatest help since at least 100 were instantly eliminated as a result of the earthquake.

Immediately after the earthquake, the city established a working group of broad representation, known as Vision Santa Cruz, whose aim was to "spearhead" the recovery process and to develop a plan that was expressive of the community's values. Significant, in terms of process and an expression of resiliency, was for Vision Santa Cruz to focus on basic principles of rebuilding. These were called the "First Principles," which established consensus over a broad spectrum of issues.

The First Principles document establishes recommended directions for a wide range of issues and topics, including the following:

1. *Form and Character of Buildings:* New buildings would be permitted to develop individual character, while retaining qualities of the historic townscape. Critical considerations: articulation, materials, signage, setbacks, scale, massing, form, bulk, solar access, and height.
2. *Building Height:* The scale and character of downtown should be maintained by two- to three-story buildings to protect the scale and character of the existing downtown, with explicit criteria for additional height, up to five stories, and provisions to ensure that buildings do not shade critical, public open spaces.
3. *Housing:* Significant opportunities for new housing would be identified throughout the downtown area, with a mixed composition of apartments, condominiums, and subsidized housing.
4. *Accessibility:* There would be an integration of aesthetic access to all improvements to ensure increased opportunities for the public to participate in commercial, governmental, residential, social, and cultural activities.
5. *Open Space and Streetscape:* A strong network of public and private open spaces (streets, sidewalks, public parks, plazas, passageways, and courtyards) would create a socially active and pedestrian-oriented downtown core.
6. *Circulation:* A predominantly pedestrian downtown would be retained; movement should be structured to reinforce the character of the place and emphasize pedestrian, bicycle, and transit access.
7. *Parking:* Parking in the downtown core should continue to be provided in a centralized fashion to maximize shared use and minimize the quantity of stored vehicles.[46]

Efforts like this help to sort through myriad variables that merit attention when contemplating significant land use actions. They do not overlook complexity but rather identify critical dimensions wherein consensus must be reached before moving forward.

Principle 9: All Relationships Are Irreversible

Because change is a constant process orchestrated along the interactive web of universal relationships, it produces infinite novelty that precludes anything in the cosmos from ever being reversible. Take my (Chris's) misstep on the aforementioned stream's edge. One moment, I was standing on the level beach, and the next I was conversing with the water. At the same time, the sand I had knocked into the stream was being summarily carried off to the sea. What of this dynamic was reversible? Nothing. Why not? Simply put, I could not go back in time and make a different decision of where to place my foot. And, because we cannot go back in time, nothing can be restored to its former condition. All we can ever do is repair something that is broken so it can continue to function, albeit differently than in its original form.

Application of Principle to Land Use Planning

Simply put, in all land use planning, the significant, irreversible changes that are anticipated as a result of alterations to the existing landscape should be identified. The consequences of these irreversible changes must also be evaluated, including potential unintended consequences. (For more information, see "Environmental Preservation, Uncertainty and Irreversibility."[47])

Example 3.9 Wetland Mitigation Banks and the No-Net-Loss Requirement

Section 404 of the Clean Water Act provides regulatory oversight for wetland conservation. A primary goal of the Section 404 program is to achieve no net loss of the remaining wetland acres (hectares) and functions. This is to be accomplished through "wetland mitigation banks." This concept has been incorporated into wetland policy because of the potential advantage in achieving the no-net-loss requirements.

A study, conducted as part of a master's thesis in 2008 relative to the application of Section 404 requirements in southeastern Louisiana, demonstrates the principle of irreversibility. The study analyzed four wetland mitigation banks with credits sold between 1991 and 2007 to determine if they were contributing to the goal of no net loss. The goal of no net loss is reached when there is complete compensation of wetland structure and function. Using data collected from permits and from bank documents on wetland size and vegetation, a structural comparison was made between wetlands in the mitigation bank and those occurring naturally. A functional assessment of the wetlands was conducted through hydrogeomorphic classification using a geographical information system to integrate information on their geology, ecoregion, slope, and soils.

The results showed that 2,546 acres (850 hectares) of mitigation were required to replace 2,309 acres (932 hectares) of impacted wetlands, creating an additional 238 acres (96 hectares) of required mitigation. "Ninety-five percent of the wetlands lost were replaced with wetlands that had similar vegetation. Furthermore, the hydrogeomorphic classification illustrated that 99 percent of all permitted wetlands were also functionally similar to their mitigated counterparts. The mitigation ratios used to calculate functional performance showed a functional loss of 2,505 acres (1,014 ha)."[48]

A positive trade was revealed when looking at wetland structure alone. However, a negative trade was revealed when looking at function. It was hypothesized that the cumulative effect of a loss in wetland functions could be a degraded water catchment that provides inferior water quality, habitat quality, water storage capacity, and so on. Thus, the mitigation requirements of Section 404 were found to be inconsistent, leading to a high functional loss for the basin.

Principle 10: All Systems Are Based on Composition, Structure, and Function

We perceive objects by means of their obvious structures or functions. Structure is the configuration of elements, parts—composition of constituents—be it simple or complex. The structure can be thought of as the organization, arrangement, or makeup of a thing. Function, on the other hand, is what a particular structure either can do or allows to be done to it or with it.

To maintain biophysical functions means that we humans must maintain the characteristics of the ecosystem in such a way that its processes are sustainable. The characteristics we must be concerned with are (1) composition, (2) structure, (3) function, and (4) nature's disturbance regimes, which periodically alter an ecosystem's composition, structure, and function.

We can, for example, change the composition of an ecosystem, such as the kinds and arrangement of plants in a forest, grassland, or agricultural crop. This alteration means that composition is malleable to human desire and thus negotiable within the context of cause and effect. In this case, composition is the determinant of the structure and function in that composition is the cause, rather than the effect, of the structure and function.

Composition determines the structure, and structure determines the function. Thus, by negotiating the composition, we simultaneously negotiate both the structure and the function. On the other hand, once the composition is in place, the structure and function are set—unless, of course, the composition is altered, at which time both the structure and function are altered accordingly.

In this sense, the composition or kinds of plants and their age classes within a plant community create a certain structure that is characteristic of the plant community at any given age. It is the structure of the plant community that in turn creates and maintains certain functions. In turn, it is the composition, structure, and function of a plant community that determine what kinds of animals can live there, how many, and for how long.

Thus, if an owner of forestland changes the composition of the forest, he or she changes the structure, hence the function, and thus affects the animals. The animals living in the forest are not just a reflection of its composition at any given point in time but are ultimately constrained by it.

If townspeople want a particular animal or group of animals within their urban growth boundary, let us say a rich diversity of summering birds and colorful butterflies to attract tourist dollars from bird-watchers and visitors in general, members of the community would have to work backward by determining what kind of function to create. To do so, they would have to know what kind of structure to create, which means knowing what type of composition is necessary to produce the required habitats for the animals the community wants. Thus, once the composition is ensconced, the structure and its attendant functions operate as an interactive unit in terms of the habitat required for the animals.

For example, the connectivity—accessibility—of habitat components is particularly important for the resident population, regardless of species, because each habitat has a biological carrying capacity, meaning a finite number of individuals that can live in a particular area without altering it to their detriment. If, however, you, the reader, either grew up in a city or now live in one, you may be familiar with habitat corridors in urban settings.

Birds in urban landscapes generally occupy parks, which are analogous to forest fragments, whereas tree-lined streets form linear corridors that connect the fragments within the urban matrix. To understand the species–habitat dynamics of an urban setting, a study conducted in Madrid, Spain, examined the effects of street location within the urbanscape, vegetative structure along the streets, and human disturbance (pedestrian and automotive) on the richness of bird species within the street corridors. In addition, the birds' temporal persistence, density of feeding and nesting guilds, and the probability of a street being occupied by a single species were also taken into account.

The number of species increased from the least-suitable habitats (streets without vegetation) to the most suitable habitats (urban parks), with tree-lined streets an intermediate landscape element. Tree-lined streets that connected urban parks influenced the number of species within the streets' vegetation positively, as well as species persistence, population density, and the probability that the individual species would continue to occupy the streets. That said, human disturbance did exert a negative influence on the same variables.

Wooded streets could potentially function as corridors that would allow certain species to fare well by supporting alternative habitats for feeding and nesting, particularly those birds that feed on the ground and nest in trees or tree cavities. Local improvements in quality and complexity of the vegetation associated with certain streets, as well as a reduction in the disturbance caused by people, could exert a positive influence on the regional connectivity of streets as a system of urban corridors for birds. Because of the differential use of corridors by species with various habitat requirements, streets as habitat corridors could be further improved by taking the requirements of different species into account.[49]

People and nature are continually altering the structure and function of this ecosystem or that ecosystem by manipulating the composition of its plants, an act that subsequently changes the composition of the animals dependent on the structure and function of the resultant habitat. By altering the composition of plants within an ecosystem, people and nature alter its structure and, in turn, affect how it functions, which in turn determines not only its potential ecosystem services but also what uses humans can derive from those ecosystem services.

Application of Principle to Land Use Planning

With the foregoing in mind, a community must plan its land use in a way that perpetuates environmental reciprocity if the community is to experience its own sustainability within that of its surrounding landscape over time.

Cavalier removal or endangerment of plant and animal species has not been uncommon, historically, in the chronicles of land use planning. Fortunately, laws now restrict these practices, in part due to the Endangered Species Act, and yet even removing invasive species, once established, can have serious consequences.

Example 3.10: Removing Invasive Species

While eradications of invasive species can often have strikingly positive effects on native biota, recent research indicates that species removal in isolation can also result in unexpected changes to other ecosystem components. Such unintended consequences "become more likely as numbers of interacting invaders increase in ecosystems, and as exotics in late stages of invasion eliminate native species and replace their functional roles." (p. 454)[50]

Principle 11: All Systems Have Cumulative Effects, Lag Periods, and Thresholds

Nature, as previously stated, has intrinsic value only and so allows each component of an ecosystem to develop its prescribed structure, carry out its biophysical function, and interact with other components through their evolved, interdependent processes and self-reinforcing feedback loops. No component is more or less important than another—except in human valuation based on personal desire for a particular outcome. Each component may differ from the other in form, but all are complementary in function.

Our intellectual challenge in land use decision making is to recognize that no given factor can be singled out as the sole cause of—or answer to—anything. All things operate synergistically as cumulative effects that exhibit a lag period before fully manifesting an outcome of some type. Cumulative effects, which encompass many little, inherent novelties, cannot be understood statistically because ecological relationships are far more complex and far less predictable than our statistical models lead us to believe—a circumstance Francis Bacon may have been alluding to when he said, "The subtlety of Nature is greater many times over than the subtlety of the senses and understanding."[51] In essence, Bacon's observation recognizes that we live in the "invisible present" and thus cannot always recognize cumulative effects.[52]

The invisible present is our inability to stand at a given point in time and see the small, seemingly innocuous effects of our actions as they accumulate over weeks, months, and years. Obviously, we can all sense change—day becoming night, night turning into day, a hot summer changing into a cold winter, and so on. But, some people who live for a long time in one place can see longer-term events and remember the winter of the exceptionally deep snow or a summer of deadly heat.

Despite such a gift, it is a rare individual who can sense, with any degree of precision, the changes that occur over the decades of their lives. At this scale of time, we tend to think of the world as being in some sort of ongoing, relatively steady state, with the exception of technology and periodic natural disasters. Moreover, we typically underestimate the degree to which slow, seemingly innocuous change has occurred—such as global warming. We are unable to sense slow changes directly, and we are even more limited in our abilities to interpret the relationships of cause and effect in these changes. To illustrate:

> The government's colorful map of planting zones is being updated for a warmer twenty-first century.
>
> The official guide for 80 million gardeners and a staple on seed packets reflects a new reality: The coldest day of the year is not as cold as it used to be. So, some plants that once seemed too vulnerable to cold can now survive farther north.
>
> It is the first time since 1990 that the U.S. Department of Agriculture has updated the planting map and much has changed. Nearly entire states, such as Ohio, Nebraska, and Texas, are in warmer zones.
>
> For example, Des Moines, Iowa, used to be in zone 5a, meaning the lowest temperature on average was between 15 and –20 degrees Fahrenheit (26 and 28.9 degrees Celsius). Now, it is 5b, which has a coldest temperature of 10 to 15 degrees below zero (23 to 26 degrees Celsius).[53]

Nevertheless, these subtle processes, acting quietly and unobtrusively over decades, reside cloaked in the invisible present, such as gradual declines or improvements in habitat quality. Thus, from a land use planning point of view, the invisible present can be fraught with the tyranny of many, little, seemingly unimportant and unrelated social-environmental decisions that invite disaster—especially when abetted by informed denial, despite the warnings of history and people with clear foresight.

For example, virtually all the world's roughly 3,000 professional climate scientists (and other professionals who work closely with them) foresee a more inhospitable world for the generations of young children already born, such as higher prices for food; growing numbers of plants and animals becoming extinct, along with their ecological functions and services; as well as other ecological disruptions linked directly to the warming climate. Yet, despite all the evidence observed and irrefutably measured worldwide, such as melting glaciers and rising ocean levels, the fossil fuel industry and ideological groups are trying to delay regulation of greenhouse emissions by funding campaigns that describe confusion and intimidation in an effort to discount the data.[54]

At length, however, cumulative effects, gathering themselves below our level of conscious awareness, suddenly become visible. By then, it is too late to retract our decisions and actions, even if the outcome they cause is decidedly negative with respect to our intentions. So it is that cumulative effects from our decisions and resultant activities can multiply unnoticed until something in the environment shifts dramatically enough—even for people without social-environmental foresight—to see the outcome through casual observation. That shift is defined by a threshold of tolerance in the system, beyond which the system as we knew it suddenly, visibly, irreversibly becomes something else. Within our world, this same dynamic takes place in a vast array of scales in all natural *and* artificial systems, from the infinitesimal to the gigantic.

A shorter-term example of cumulative effects, lag period, and threshold is the cutting down of a dying tree along a city street. Initially, a man from the tree service saws off the small branches with intact twigs. The effect is barely discernible at first, even as the branches begin to pile up on the ground. Each severed branch represents a cumulative effect that would be all but unnoticeable if they did not accumulating under the tree.

After an hour or so (lag period) of removing the small limbs on one side of the tree, the cumulative effects gradually become visible as they cross a structurally irreversible threshold. Had the same volume of twigs been removed from throughout the tree and simultaneously gathered and removed from the ground, the cumulative effects would not have been as apparent. Nevertheless, the tree is gradually transformed into a stark skeleton of larger branches and the main trunk. Then, the large branches are cut off a section at a time, with the same visual effect as when the small ones were removed, until only the trunk remains. The piecemeal removal of the tree creates a slowly changing vista until an opening in the street's corridor of trees presents, for the first time in decades, an unobstructed view of the buildings beyond where the tree once stood as a stark threshold is crossed.

And so the process of change and novelty continues unabated in all its myriad and astounding scales because of economic/political "informed denial." Here is an ideal situation to point out the dynamics of informed denial: In response to global warming and rising sea levels, which are submerging deltas and oceanic islands and slowly inundating such coastal cities as New Orleans, the engineering solution—based on the U.S. Army Corps of Engineers' symptomatic decisions—has been to disregard the ongoing change and build higher levees, a patently futile response in the long-term effects of global warming and its oceanic consequences.

Application of Principle to Land Use Planning

How can a community account for the cumulative effects, lag periods, and thresholds of its land use planning? The precautionary principle comes to mind. This principle

> allows policy makers to make discretionary decisions in situations where there is the possibility of harm from taking a particular course or making a certain decision when extensive scientific knowledge on the matter is lacking. The principle implies that there is a social responsibility to protect the public from exposure to harm, when scientific investigation has found a plausible risk. These protections can be relaxed only if further scientific findings emerge that provide sound evidence that no harm will result.[55]

Example 3.11: Tasmania—Land Use Planning and Approvals Act of 1993

The Land Use Planning and Approvals Act of 1993 is the primary planning legislation in Tasmania. Its purpose is to promote ecologically sustainable development to protect and enhance Tasmania's resources for the benefit of future generations.

The policy is based on two major and increasingly dominant international and national trends in decision making. The first is the movement toward greater community participation in decision making, including in the policy, planning, and management stages. The second is the growing importance and recognition of the precautionary principle in decision making. In the language of the act: "The precautionary principle is widely recognized internationally and nationally as a key factor in achieving ecologically sustainable development and as a key component of successful environmental management and planning. ... The precautionary principle should be applied to land-use planning and development at both a macro and micro scale."[56]

Principle 12: All Systems Are Cyclical, But None Is a Perfect Circle

While all processes in nature are cyclical, no cycle is a perfect circle, despite such depictions in the scientific literature and textbooks. They are, instead, a coming together in time and space at a specific point, where one "end" of a cycle approximates—but only approximates—its "beginning" in a particular time and place. Between its beginning and its ending, a cycle can have any configuration of cosmic happenstance. Clearly, there will be change, but the degree and form of the change will be determined by the "original" conditions. Biophysical cycles can thus be likened to a coiled spring insofar as every coil approximates the curvature of its neighbor but always on a different spatial level (temporal level in nature), thus never touching.

Regardless of its size or flexibility, a spring's coils are forever reaching outward. With respect to nature's biophysical cycles, they are forever moving toward the next level of novelty in the creative process and so are perpetually embracing the uncertainty of future conditions—never to repeat the exact outcome of an event as it once happened. This phenomenon occurs even in times of relative climatic stability. Be that as it may, today's progressive global warming will only intensify tomorrow's uncertainties—such as the potential outcomes of land use planning.

In human terms, life is composed of rhythms or routines that follow the cycles from the minute to the infinite. We humans most commonly experience the nature of cycles in our pilgrimage through the days, months, and years of our lives wherein certain events are repetitive—day and night, the waxing and waning of the moon, the march of the seasons, and the coming and going of birthdays, all marking the circular passage we perceive as time within the curvature of space—although many may perceive life as a linear progression from birth to death. In addition to the visible manifestation of these repetitive cycles, nature's biophysical processes are cyclical in various scales of time and space, a phenomenon that means all relationships are simultaneously cyclical in their outworking and forever novel in their outcomes. Regarding cycles, farmer and author Wendell Berry said, "It is only in the processes of the natural world, and in analogous and related processes of human culture, that the new may grow usefully old, and the old be made new."[57]

Application of Principle to Land Use Planning

This principle relates strongly to principle 8—change is a process of eternal becoming—in that it urges us not only to accept that all systems are cyclical, as opposed to perfect circles, but also to evaluate where the systems in a proposed development are in the cycle of change. This means natural systems as well as social systems.

Example 3.12: Replacing a Shopping Center with an Ecological Neighborhood, Phalen Park, St. Paul, Minnesota, 2005

The idea of replacing one of the oldest shopping centers in St. Paul, Minnesota, with a wetland park was first proposed by landscape architects Joan Nassauer, Harrison Fraker, and students at the University of Minnesota. Their task was to assist the city of St. Paul and the residents around Phalen Lake in exploring a range of design alternatives for a deteriorating and financially bankrupt shopping center.

This design recognized the existence of a wetland substructure on which the shopping center had been built. A small lake had existed before the site had been filled. The wetland "park" concept became the centerpiece of a vision adopted as part of a city-funded Phalen Village Small Area Plan—a transportation-oriented, mixed-use development, including a significant amount of affordable housing.

The city began negotiations to purchase the shopping center in 1994, and by 1998 demolition had begun to create the new Phalen Wetland Park. In addition to "creating a signature public amenity, the plan envisioned the park linking to a new wildlife habitat 'spine' and helping filter and retain local storm water." The first phase of the park, on land owned by the city, was constructed in 1997, and the second phase began in 1998. Development, including new businesses, institutions, and housing according to the Small Area Plan, began in 1998.

As the plan unfolded and development ensued, however, rather than maintaining crisp boundaries between the park and adjacent commercial and residential areas, much of the new residential development had been arrayed around the wetland on small cul-de-sacs. "This has brought a miniature version of a familiar suburban scene—wetlands in the backyards of new single-family homes—into the heart of one of the oldest, most urban neighborhoods of St. Paul."[58]

The good news was that construction of the urban ecosystem, which emulated the natural system, seemed to work. Studies indicated that the re-created wetlands, constructed during phase 1, dramatically increased local biodiversity.

The bad news was that new housing developments encroached on the planned boundaries of the wetland park itself. While all the housing was a welcome addition to the neighborhood, there was "no doubt its location … severely compromised the original purpose of the wetland park as a public amenity that embodied urban ecological function. In the original plan, the wetland park's boundary was sharply defined by public streets, making it both visible and accessible as the centerpiece of development. Now, with the intrusion of housing into it, the park is no longer experienced as 'open' and available to the public. It is now defined by private property."[59]

The Phalen Wetland Park story brings to the fore how private, suburban development can overwhelm the identity and definition of public spaces as developers—recognizing their monetary value—incorporate them into their private projects to increase the projects' value.[60]

Principle 13: Systemic Change Is Based on Self-Organized Criticality

When dealing with scale (a small mountain lake as opposed to the drainage basin of a large river, such as the Mississippi in the United States or the

Ganges in India), scientists have traditionally analyzed large, interactive systems in the same way that they have studied small, orderly systems, mainly because their methods of study had proven so successful to their sense of understanding. The prevailing wisdom has been that the behavior of a large, complicated system could be predicted by studying its elements separately and by analyzing its microscopic mechanisms individually—such reductionist-mechanical thinking is predominant in Western society and tends to view the world and all it contains through a lens of intellectual isolation ensconced in symptomatology. During the last few decades, however, it has become increasingly clear that many complicated systems, like forests, oceans, and even cities, do not yield to such traditional analysis.

Instead, large, complicated, interactive systems seem to evolve naturally to a critical state in which even a minor event starts a chain reaction that can affect any number of internal elements and can lead to a dramatic alteration in the system. Although such systems produce more minor events than catastrophic ones, chain reactions of all sizes are an integral part of system dynamics. According to the theory called *self-organized criticality*, the mechanism that leads to minor events (analogous to the drop of a pin) is the same mechanism that leads to major events (analogous to an earthquake).[61] With respect to earthquakes, for example, the weight of seasonal rains or snow on the soil, or floodwaters, such as might collect in a lake—and by analogy a reservoir—in areas of geological faults can be heavy enough to slightly depress a fault zone and thus cause initial movement along the fault.[62]

Thus, every ecosystem moves inevitably toward a critical state that alters it in some dramatic way. This dynamic makes ecosystems dissipative structures in that energy is built up through time only to be released in a disturbance of some kind, such as a fire, flood, or landslide in some scale, ranging from a freshet in a stream to the eruption of a volcano, after which energy begins building again toward the next release of pent-up energy somewhere in time.

These "disturbances," as ecologists think of these events, can be long term and chronic, such as large movements of soil that take place over hundreds of years (termed an *earth flow*), or acute, such as the crescendo of a volcanic eruption that sends a pyroclastic flow sweeping down its side at amazing speed. (A *pyroclastic flow* is a turbulent mixture of hot gas and fragments of rock, such as pumice, that is violently ejected from a fissure and moves with great speed down the side of a volcano. *Pyroclastic* is Greek for "fire-broken.")

Here, you might interject that neither a movement of soil nor a volcano is a living system in the classical sense. Although true, all disturbance regimes are part and parcel of the living systems they affect. Thus, interactive systems, from the habitat of a gnat to a tropical rain forest, perpetually organize themselves to a critical state in which a minor event can start a chain reaction that leads to a catastrophic event—as far as living things (such as humans) are concerned—after which the system begins organizing itself toward the next critical state. Therefore, systems never reach a state of equilibrium but rather evolve continuously from one semistable state to another.

Application of Principle to Land Use Planning

This dynamic is precisely why the sustainability of land use planning is a moving target for decision makers—not a fixed end point or a steady state, as intimated in so-called steady-state economy and closed-loop technology, both biophysical impossibilities. Principle 8 suffices as an example of how principle 13 fits into land use planning.

Principle 14: Dynamic Disequilibrium Rules All Systems

If change is a universal constant in which nothing is static, what, then, is a natural state? In answering this question, it becomes apparent that the "balance of nature" in the classical sense (disturb nature and nature will return to its former state after the disturbance is removed) is a fond imagining. In fact, the so-called balance of nature is a romanticized figment of the human imagination, something we conjured to fit our hoped-for, snapshot image of the world in which we live. In reality, nature exists in a continual state of ever-shifting *disequilibrium*: Ecosystems are entrained in the irreversible process of change and novelty, thereby altering their composition, structure, function, and the resultant interactive feedback loops in which the available resources are used—irrespective of human influence. Perhaps the most outstanding evidence that an ecosystem is subject to constant change and disruption, rather than remaining in a static balance, comes from studies of naturally occurring external factors that dislocate ecosystems, and climate appears to be foremost among these factors.

After a fire, earthquake, volcanic eruption, flood, hurricane, or landslide, for example, a biological system may eventually be able to approximate what it was through resilience—the ability of the system to retain the integrity of its basic relationships. In this way, an old forest that is burned, blown over in a hurricane, or smashed in a tsunami can be replaced by another—albeit different—old forest on the same acreage (number of hectares). Thus, despite a repetitive disturbance regime, a forest ecosystem can remain a forest ecosystem, although much changed in composition, structure, and function.[63] So it is that ancient forests around the world have been evolving from one critical, biophysical state to the next, from one natural catastrophe to the next through a process of what is termed *autogenic* (produced from within, self-generating) succession. Autogenic succession can be characterized by "successional stages," a concept that refers to the characteristic developmental stages that a forest goes through from bare ground to an ancient or "old-growth" forest.

Autogenic succession works as follows: Grasses and other herbaceous plants are the first to grow in a burned area and so constitute the first successional stage following a fire. In their occupation of a given site and in their growing, they gradually alter the characteristics of the soil, such as pH, until it is no longer optimum for their survival and growth. Their offspring may germinate but not survive, creating areas occupied only by parent plants. As the offspring succumb to the changes in the soil and the parent plants age, die, and are not replaced, openings appear in the vegetative cover that allow shrubs to become established.

In this way, the herbaceous stage changes to an early shrub stage. In turn, the shrub stage follows the same process as the herbaceous stage and gives way to tree seedlings that, in combination, become the shrub-seedling stage. As the shrubs die, the shrub-seedling stage gives way to the sapling stage, which grows into a young forest stage, a mature forest stage, and finally an old-growth forest stage—that is, until succession starts again. The six generalized autogenic, successional stages that a Western coniferous forest goes through can be characterized as follows: herbaceous → shrub-seedling → sapling → young forest → mature forest → old-growth forest → fire or other disturbance, which starts the cycle again and eventually again. However, rather than "discrete" stages of development that precede in an orderly succession, one after another, the "stages" of forest development form a complex continuum in which each stage builds on the dynamics and biophysical nuances of the preceding one. Therefore, no forested areas ever develop alike in an absolutely repetitive fashion.[64]

But, regardless of how closely an ecosystem might approximate its former state following a disturbance, the existence of every ecosystem is a tenuous balancing act because every system is in a continual state of reorganization that occurs over various scales of time, from the cycle of an old forest to a geological phenomenon, such as Mona Loa, the active volcanic mountain in Hawaii. Whereas people can manipulate a forest to some extent, Mona Loa is entrained in an eternal flux of physical novelty over which no human has a smidgen of control.

Application of Principle to Land Use Planning

Principle 8 suffices as an example of how Principle 14 fits into land use planning.

Closing Comment

Finally, we, the human component of the world, must understand and accept that the foregoing biophysical principles are an interactive thread in the tapestry of the natural/cultural world that must be accounted for—and honored—in land use planning if society is to become a sustainable partner with its various environments. As such, they are an essential and unavoidable part of the "commons," which includes such things as clean air, pure water, fertile soil, viable forests, and bountiful oceans as the birthright of every human being and life in general—the "global commons." Protecting the long-term biophysical viability of the global commons through local, regional, national, and global land use planning is the foremost responsibility of every decision maker—regardless of the nomenclature by which they are known.

Summary

Chapter 3 is a discussion of the inviolable biophysical principles that form the nonnegotiable constraints whereby humans can practice conscious, purposeful social-environmental sustainability to their benefit. When, however, those limits are ignored, environmental degradation ensues, along with a diminished productive capacity of those ecosystems on which we rely for a good quality of life—a situation that is not only progressively irreversible but also passed forward to all generations. Chapter 4 is a critical appraisal of what it takes to plan for a local living economy in the form of social-environmental sustainability by reinventing the comprehensive plan.

Endnotes

1. T. S. Eliot. http://www.brainyquote.com/quotes/quotes/t/tseliot109032.html (accessed July 18, 2012).
2. Marie Curie. http://thinkexist.com/quotation/nothing_in_life_is_to_be_feared-it_is_only_to_be/14155.html (accessed July 20, 2012).
3. Two examples are (1) "City of Seattle Sustainable Design Checklist." May 9, 2006. http://www.seattle.gov/dpd/cms/groups/pan/@pan/@sustainable-blding/documents/web_informational/dpds_007169.pdf (accessed December 1, 2012); and (2) "Environmental Information Document Review Checklist." n.d. http://www.epa.gov/region9/water/grants/sap/pdfs/EID-checklist.pdf (accessed December 1, 2012).
4. Indigo Development. "The Industrial Symbiosis at Kalundborg, Denmark." 2003. http://www.indigodev.com/Kal.html (accessed November 26, 2012).
5. Ibid.
6. Ibid.

7. Jonathan Hiskes. "The Case for Super-Ambitious Living Buildings. A Talk with Jason McLennan." 2010. http://grist.org/article/2010-09-30-jason-mclennan-living-buildings-interview/ (accessed December 8, 2012).
8. Indigo Development, "Industrial Symbiosis."
9. Russ Beaton and Chris Maser. *Economics and Ecology: United for a Sustainable World*. CRC Press, Boca Raton, FL, 2011.
10. Chris Maser. *Earth in Our Care: Ecology, Economy, and Sustainability*. Rutgers University Press, Piscataway, NJ, 2009.
11. Open Politics. "Consensus Decision Making." n.d. http://openpolitics.ca/consensus+decision+making (accessed March 8, 2013).
12. Slow Movement. "The Slow Movement: Making a Connection." n.d. http://www.slowmovement.com/ (accessed November 25, 2012).
13. Cittaslow International Network home page. 2011. http://www.cittaslow.org/ (accessed December 17, 2012).
14. Slow Food USA. "What Is Slow Food." 2010. http://www.slowfoodusa.org/index.php/slow_food/ (accessed December 17, 2012).
15. (1) Yuan-Chug Cheng and Graham R. Fleming. "Dynamics of Light Harvesting in Photosynthesis." *Annual Review of Physical Chemistry* 60 (2009):241–262; (2) Paul May. "Chlorophyll." http://www.chm.bris.ac.uk/3motm/chlorophyll/chlorophyll_h.htm (accessed January 5, 2009); (3) T. R. Calhoun, N. S. Ginsberg, G. S. Schlau-Cohen, and others. "Quantum Coherence Enabled Determination of the Energy Landscape in Light-Harvesting Complex II." *Journal of Physical Chemistry B* 113 (2009):16291–16295; and (4) Susan Gaidos. "Living Physics." *Science News* 175(10) (2009):24–29.
16. Solar America Board for Codes and Standards. "A Comprehensive Review of Solar Access Law in the United States." n.d. http://www.solarabcs.org/about/publications/reports/solar-access/pdfs/SolarAccess.pdf (accessed December 18, 2012).
17. Ibid.
18. American Planning Association. "Solar Access Ordinances." Policy Guides. 2007. http://www.planning.org/pas/infopackets/open/pdf/30part3.pdf (accessed November 29, 2012).
19. James Fennimore Cooper. *The Last of the Mohicans,* Bantam Classic edition. Bantam Dell, New York, 1981.
20. The foregoing three paragraphs are based on Minnesota Pollution Control Agency. "Sample Criteria for Evaluating the Sustainability of Community Ideas and Projects." 2010. http://www.pca.state.mn.us/index.php?option=com_k2&view=item&id=2319 (accessed December 18, 2012).
21. Clinton D. Francis, Nathan J. Kleist, Catherine P. Ortega, and Alexander Cruz. "Noise Pollution Alters Ecological Services: Enhanced Pollination and Disrupted Seed Dispersal." *Proceedings of the Royal Society B* 279 (2012):2727–2735.
22. Elizabeth Kline. "Why Sustainable Community Indicators? People Need a Reality Check." *Wingspread Journal,* 18(2) (Spring 1996). http://www.smartcommunities.ncat.org/articles/whysust.shtml (accessed December 17, 2012).
23. Sustainable Seattle. "Indicators of Sustainable Community." 1998. http://docs.google.com/a/peak.org/viewer?a=v&q=cache:pVv8Fr6M0wEJ:ftp.utalca.cl/redcauquenes/cauquenes%2520estudio/Articulos/SEATTLE_SUSTAINABLE%2520SEATTLE%2520INDICATORS.pdf+http://www.sustainableseattle.org/1998IndicatorsReport.pdf&hl=en&gl=us&pid=bl&srcid=AD

GEESilRMWSmopCX5nKdQe1OpQ-Lu1wbDHmc3mjxqqlupoz7sg17dF6A-SoED0l_RA0ROM_GgwwSWG0iRjW_TE8EMxycNuUpAgJXf_eHhJ7bN-2paBYM-0HbDT5kuJNeMSfBN7uV8EuvG&sig=AHIEtbSQSgEOj_2jIfRPZ4E5kEA3-FWtPw (accessed October 30, 2012).

24. Ibid.
25. Maser, *Earth in Our Care*.
26. The following discussion is drawn from Jane Silberstein. "Expanding Prosperity by Becoming an Eco-municipality." *Business and Society Review* 115 (2010):465–475.
27. Ibid.
28. Sierra Business Council. "Investing for Prosperity." 2003. 144 pp. http://www.sbcouncil.org/pdf/Investing%20for%20Prosperity.pdf (accessed September 21, 2012).
29. NextStep. "Eco-Industrial Project in Superior, WI." 2011. http://www.nextstep.state.mn.us/res_detail.cfm?id=2206 (accessed November 6, 2012).
30. William McDonough and Michael Braungart. *Cradle to Cradle: Remaking the Way We Make Things*. Farrar, Strauss and Giroux, New York, 2002.
31. Fritjof Capra. *Turning Point: Science, Society and the Rising Culture*. Bantam Books, New York, 1983.
32. Sustainable Lawrence. "The Eco-municipality Model for Sustainable Community Change: A Systems Approach to Creating Sustainable Communities." 2006. http://www.sustainablelawrence.org/ecomunimodel.pdf (accessed December 1, 2012).
33. Peter M. Senge, Joe Laur, Sara Schley, Nina Kruschwitz, and Bryan Smith. *The Necessary Revolution: How Individuals and Organizations Are Working Together to Create a Sustainable World*. Doubleday, New York, 2008.
34. (1) Sid Perkins. "As Ozone Hole Heals, Antarctic Could Heat Up." *Science News* July 5, 2008:10; (2) S.-W. Son, L. M. Polvani, D. W. Waugh, and others. "The Impact of Stratospheric Ozone Recovery on the Southern Hemisphere Westerly Jet." *Science* 320 (2008):1486–1489; and (3) J. Perlwitz, S. Pawson, R. L. Fogt, and others. "Impact of Stratospheric Ozone Hole Recovery on Antarctic Climate." *Geophysical Research Letters* 35 (2008) L08714, doi:10.1029/2008GL033317 (accessed December 17, 2011).
35. http://www.slcgov.com/slcgreen/urbanforestry/ (accessed July 7, 2013).
36. http://www.slcdocs.com/slcgreen/sustainable%20SLC2015_AirQuality+ClimateChange.pdf (accessed July 7, 2013).
37. The discussion under item 3 is based on "Community Gardening and Urban Farming." http://www.slcgov.com/slc-green/communitygarden (accessed December 19, 2012).
38. (1) Maser, *Earth in Our Care*; and (2) U.S. Geological Survey. *A Brief History and Summary of the Effects of River Engineering and Dams on the Mississippi River System and Delta*. USGS, Reston, VA, 2012. http://pubs.usgs.gov/circ/1375/C1375.pdf (accessed December 19, 2012).
39. Lee Dye. "Did Climate Change Kill the Roman Empire?" http://abcnews.go.com/Technology/JustOneThing/story?id=6428550&page=1 (accessed December 10, 2010).
40. (1) Sid Perkins. "Buried Treasures." *Science News*, 169 (2006):266–268; (2) Martin B. Short, James C. Baygents, and Raymond E. Goldstein. "Stalactite Growth as a Free-Boundary Problem." *Physics of Fluids* 17 (2005):083101, 12 pp. http://

dx.doi.org/10.1063/1.2006027 (accessed December 17, 2008); and (3) M. B. Short, J. C. Baygents, J. W. Beck, and others. "Stalactite Growth as a Free-Boundary Problem: A Geometric Law and Its Platonic Ideal." *Physical Review Letters* 94 (2005):018510, 4 pp. http://link.aps.org/doi/10.1103/PhysRevLett.94.018501 (accessed December 17, 2008).

41. (1) Ian J. Orland, Miryam Bar-Matthews, Noriko T. Kita, and others. "Climate Deterioration in the Eastern Mediterranean as Revealed by Ion Microprobe Analysis of a Speleothem that Grew from 2.2 to 0.9 Ka in Soreq Cave, Israel." *Quaternary Research* 71 (2009):27–35; (2) A. Kaufman, G.J. Wasserburg, D. Porcelli, and others. "U-Th Isotope Systematics from the Soreq Cave, Israel and Climatic Correlations." *Earth and Planetary Science Letters* 156 (1998):141–155; and (3) Avner Ayalon, Miryam Bar-Matthews, and Eytan Sass. "Rainfall-Recharge Relationships within a Karstic Terrain in the Eastern Mediterranean Semi-Arid Region, Israel: $\delta\ ^{18}O$ and δD Characteristics." *Journal of Hydrology* 207 (1998):18–31.
42. (1) Jed O. Kaplan, Kristen M. Krumhardt, Erle C. Ellis, and others. "Holocene Carbon Emissions as a Result of Anthropogenic Land Cover Change." *The Holocene* 21 (2011):775–791; (2) Jonathan A. Foley, Chad Monfreda, Navin Ramankutty, and David Zaks. "Our Share of the Planetary Pie." *Proceedings of the National Academy of Sciences* 104 (2007):12585–12586.
43. (1) Pingzhong Zhang, Hai Cheng, R. Lawrence Edwards, and others. "A Test of Climate, Sun, and Culture Relationships from an 1810-Year Chinese Cave Record." *Science* 322 (2008):940–942; (2) Kallie Szczepanski. "When the Rains Stop, the Emperors Fall." n.d. http://asianhistory.about.com/od/asianenvironmentalhistory/a/ChinaMonsoon.htm (accessed January 5, 2009); (3) Ker Than. "Chinese Kingdoms Rose, Fell with Monsoons?" *National Geographic News*. 2008. http://news.nationalgeographic.com/news/2008/11/081106-monsoons-china.html (accessed January 10, 2009); and (4) Yongjin Wang, Hai Cheng, R. Lawrence Edwards, and others. "Millennial- and Orbital-Scale Changes in the East Asian Monsoon over the Past 224,000 Years." *Nature* 451 (2008):1090–1093.
44. Charles Darwin. *On the Origin of Species*. Modern Library, Random House, New York, 1998.
45. Karina Rusk. "Downtown Santa Cruz in Ruins after Quake." KGO-TV/DT. October 17, 2009. http://abclocal.go.com/kgo/story?section=news/local&id=7048962 (accessed December 5, 2012).
46. City of Santa Cruz. "Downtown Recovery Plan." 1991. http://www.cityofsantacruz.com/Modules/ShowDocument.aspx?documentid=8911 (accessed December 18, 2012).
47. K. J. Arrow and A. C. Fischer. "Environmental Preservation, Uncertainty and Irreversibility." *Quarterly Journal of Economics* 88 (1974):312–319.
48. Abbey Tyrna. "Wetland Mitigation Banks and the No-Net-Loss Requirement: An Evaluation of the Section 404 Permit Program in Southeast Louisiana." MS thesis. 2008. http://etd.lsu.edu/docs/available/etd-04102008-141642/unrestricted/Tyrna_thesisx.pdf (accessed November 28, 2012).
49. The three previous paragraphs are based on Esteban Fernández-Juricic. "Avifaunal Use of Wooded Streets in an Urban Landscape." *Conservation Biology* 14 (2000):513–521.

50. Erika S. Zavaleta, Richard J. Hobbs, and Harold A. Mooney. "Viewing Invasive Species Removal in a Whole-Ecosystem Context." *Trends in Ecology and Evolution* 16 (2001):454–459.
51. Francis Bacon. http://Science.prodos.ORG (accessed January 2, 2009).
52. John J. Magnuson. "Long-Term Ecological Research and the Invisible Present." *BioScience* 40 (1990):495–501.
53. Seth Borenstein. "New Federal Map for What to Plant Reflects Warming." http://abcnews.go.com/Technology/wireStory/federal-map-plant-reflects-warming-15438721#.TyBJcBw0i4A (accessed January 25, 2012).
54. Bill Blakemore. "Shakespeare, Global Warming, Sunset, and You." November 5, 2011. http://abcnews.go.com/blogs/technology/2011/11/shakespeare-global-warming-sunset-and-you/ (accessed November 7, 2011).
55. "Precautionary Principle." n.d. http://en.wikipedia.org/wiki/Precautionary_principle (accessed September 20, 2012).
56. Environmental Defenders Office. "National Competition Policy Review of Land Use Planning and Approvals Act 1993." October 1999. http://www.edo.org.au/edotas/submissions/lupaa.html (accessed September 30, 2012).
57. Wendell Berry. "The Road and the Wheel." *Earth Ethics* 1 (1990):8–9.
58. The foregoing discussion of Phalen Park, St. Paul, Minnesota, is drawn from Jennifer Dowdell, Harrison Fraker, and Joan Nassauer. "Replacing a Shopping Center with an Ecological Neighborhood." n.d. http://designobserver.com/media/pdf/Replacing_a_Sh_396.pdf (accessed August 23, 2012).
59. Ibid.
60. (1) Dowdell et al., "Replacing a Shopping Center"; and (2) I. Nassauer. "Monitoring the Success of Metropolitan Wetland Restorations: Cultural Sustainability and Ecological Function." *Wetlands* 24 (2004):756–65.
61. Per Bak and Kan Chen. "Self-Organizing Criticality." *Scientific American* January (1991):46–53.
62. (1) Devin Powell and Alexandra Witze. "Weather Affects Geologic Activity." *Science News* 180 (2011):8; (2) Daniel Brothers, Debi Kilb, Karen Luttrell, and others. "Loading of the San Andreas Fault by Flood-Induced Rupture of Faults Beneath the Salton Sea." *Nature Geoscience* 4 (2011):486–492; and (3) Thomas J. Sheeran. "Expert: Wastewater Well in Ohio Triggered Quakes." http://abcnews.go.com/US/wireStory/expert-wastewater-ohio-triggered-quakes-15275844#.TwJKDhw0gpc (accessed January 3, 2012).
63. (1) Peter H. Morrison and Frederick J. Swanson. *Fire History and Pattern in a Cascade Range Landscape*. USDA Forest Service General Technical Report. PNW-GTR-254. Pacific Northwest Research Station, Portland, OR, 1990, 77 pp.; and (2) Charles Grier Johnson, Jr. *Vegetation Response after Wildfires in National Forests of Northeastern Oregon*. R6-NR-ECOL-TP-06-98. USDA Forest Service, Pacific Northwest Region, Portland, OR, 1998, 128 pp.
64. The discussion of autogenic succession is based on Chris Maser. *Our Forest Legacy: Today's Decisions, Tomorrow's Consequences*. Maisonneuve Press, Washington, DC, 2005.

4

Planning for a Local Living Economy: Reinventing the Comprehensive Plan

> Nature's economy is the first economy, the primary economy on which all other economies rest.
>
> —**Vandana Shiva**[1]

Living Economy Defined

In designing a new path to planetary health and well-being, economist David Korten and Vandana Shiva, environmental activist, and formerly one of India's leading physicists, both refer to the need to transform our current economic system. They describe the virtues and necessity of a "living economy."[2] In such a new economy, both prosperity and poverty would be redefined since the measure of wealth is no longer confined to property, possessions, and accumulation of financial capital. "Because the most important forms of real wealth are living wealth," explains Korten, "the term *living economy* is a synonym for real-wealth economy that mimics the organization of a healthy ecosystem. The measure of a living economy's wealth is the vitality or creative life energy embodied in its people, relationships, and natural environment" (p. 109).[3]

Vandana Shiva explains a "living economy" as an essential part of an "earth democracy," where "every being has equal access to the earth's resources that make life possible; this access is assured by recognizing the importance of the other two economies: nature's economy and the sustenance economy" (p. 13).[4] To plan for a living economy, it is critical that we understand our communities as living *systems* where everything is in relationship with everything else, and no part can be viewed in isolation from the other parts.

Community as Ecosytem

Most of us want to think of our community as our home or a place we want to feel at home. When we feel at home in our communities, we understand what sense of place means. We feel welcome, safe, perhaps even needed and know that our basic requirements will be met. To plan for our communities in a way that retains that which we cherish most, incorporates the well-being of future generations, and embraces a radical course correction so needed at this time, a holistic systems approach aligned with biophysical principles is ours to choose. This approach would also be consistent with what Paul Hawken refers to as "the first rule of sustainability" which "is to align with natural forces, or at least not try to defy them."[5]

As noted in a previous chapter, ecology and economics have the same root: *eco* from the Greek *oikos*, or "home." Ecology is the knowledge or understanding of the house, and economics is the management of the house—and it is the same house. Therefore, understanding our community as a living system—an ecosystem—not only will give us a new understanding of "economy" and "economics" but also will foster a vision of the future along with strategies for its realization that equate with resiliency, adaptability, and attunement with nature.

If we perceive ourselves and all we create as part of a living ecosystem, we will then be required to rethink what we do and how we do it: from planning for our cities and towns, to government, food production, communications, waste management, transportation, and so on. The "reward," to use the term coined by Ecotrust, is "reliable prosperity"[6] or an economy that "inherently serves the self-interest of individuals, communities, and nature," and therefore fully supports life, happiness, and well-being for the long term.

An ecosystem, by definition, is a community of living organisms that, in conjunction with the nonliving components of their environment, interact as a system. These components are linked through nutrient cycles and energy flows.[7] Within this context, the economy of a community remains the composite of its wealth and resources, and the nature of "wealth" expands beyond financial capital to include both environmental and social capital in all forms. Community wealth is therefore dependent on the strength and health of networks supporting energy flow and nutrient cycles that sustain these various forms of capital. As a result, community economics is about the flow of energy. Energy, in this context, includes intelligence, information, money, knowledge, networks, and other forms of human capital, natural resources, and all ecosystem services.

Thus, the wealth of a local economy is determined by the health of the networks that provide for the smooth transfer of energy between nodes of community capital: human, financial, and natural. Good trusteeship of the sources and flow of life-giving energy means we reinvest; we monitor our use; and we make conscious choices in how we define and create wealth, as well as how to use, convert, and transfer that wealth throughout the social-environmental system.

Is there a free flow of and equity in how financial and natural resources are distributed in society? Are energy, food, health care, and education available and accessible to all? Do we reinvest in the earth and all of its resources at a rate equal to the rate of extraction?

The purpose of such a "new economy," according to David Korten, is to "support the healthy development and function of people, families, community, and nature ... and economic performance is properly assessed against these outcomes" (p. 261).[8] The "rules of the house" in such a system are therefore the nonnegotiable biophysical principles and elements of sustainability that rest on those principles—both set forth in previous chapters. One could, therefore, think of such a new economy as an institution of "applied ecology," as suggested by Korten.[9]

In a living economy, both poverty and prosperity take on new meaning. For example, not having basic requirements met, such as food, shelter, water, clothing, and privacy, would still equate with varying levels of poverty. Manfred Max-Neef and others propose a taxonomy of human necessities and a process by which communities can identify their "wealth" and "poverty" according to how their fundamental human needs are satisfied. The fundamental requirements he identifies are subsistence, protection, affection, understanding, participation, leisure, creation, identity, and freedom.[10] In a living economy, prosperity would be the satisfaction of the human needs described by Max-Neef and would likely be associated with family, connection with one another and nature, and what Bill McKibben calls a sense of "enoughness."[11] Poverty, on the other hand, would be the absence of such sources of happiness and well-being in one's life.

With a new meaning of prosperity, the inherent rights and value of all life would be honored and nurtured; human creativity would thrive, and we would have definitely made a critical and long-awaited course correction.

Comprehensive Plan for a Local Living Economy

Local governments are the primary regulators of land use in the United States. Although state and federal laws may take precedence in certain cases, such as landfills, industrial effluent, and the like, the local government serves as the intermediary for such regulations and has the responsibility to ensure compliance. Local zoning ordinances contain land use regulations that emerge from, or are tied closely to, land use goals, objectives, and actions found in the municipality's comprehensive or general plan.

These plans typically contain a variety of elements reflecting areas of community life over which the government has regulatory or policy authority. At a minimum, the following elements are usually present: housing, transportation, natural resources, economic development, and land use. In some states, like California, the elements of comprehensive municipal plans are mandated. And, in some cases, cities, like Davis, California, develop and adopt traditional comprehensive plans with the state-mandated elements but, in addition, develop a separate "sustainability plan."[12]

The elements of a comprehensive plan are intended to represent an understandable and rational division of community requirements. Each element ideally includes a vision statement, goals, objectives, and strategies for implementation. The alternative framework we are proposing is one in which the elements are closely tied to community values and a wider array of perceived necessities for both the human and nonhuman communities. Such a structure is intended to be dynamic and organic, with *organic* defined as "a complex structure of interdependent and subordinate elements whose relations and properties are largely determined by their function in the whole."[13]

As such, we do our best to demonstrate how a sustainability plan and a comprehensive plan can be essentially the same plan. We start by recasting community as an ecosystem calls forth a new format for a comprehensive plan. This reconceptualization of community enables us to view land use, as well as all other components of the plan, as part of a system wherein every part is in relationship with all other parts.

When we think of our community as a living system, the "elements" of a comprehensive plan become the centers of wealth or capital of a local living economy. These centers of wealth, or "nodes," interact as a system and are linked through nutrient cycles and energy/material flows. Thus, each node in this living system receives both deposits and withdrawals, so to speak, from the other nodes; therefore, each node is the source of wealth/energy that converts such wealth/energy and transfers wealth/energy. In this framework, all nodes benefit from the free flow of the energy, and as a result, a new understanding of economic integrity emerges: a network enabling the sustainable flow of energy and capital—human, financial, and natural. In such a system, just as in ecosystems, energy is constantly brought in, broken down or converted, and used, not only for repair but also for regeneration and to adapt to changing environmental conditions.[14]

The community, therefore, is an economic system comprised of three main forms of capital: human, financial, and natural. It follows, therefore, that a comprehensive plan could reasonably be comprised of these three elements, a structure within which all functions controlled by or strongly influenced by local government are addressed by stated goals and strategies.

For each of the following "elements" of the comprehensive plan for a local living economy, we identify what government can do to contribute to the vitality and integrity of each. We have sought examples that represent a "fit" with the principles, concepts, and values of a local living economy. These examples therefore represent efforts or initiatives that acknowledge and seek to align with biophysical principles by (1) achieving zero waste; (2) maintaining or contributing to biodiversity; and (3) making contributions to human, financial, and natural capital, thus strengthening the flow of energy within the living system.

Human Capital

The purpose of the human capital element of the local living economy comprehensive plan is to indicate how local government will support, enhance, and maintain the health and well-being of its residents while building community resiliency.

Human capital represents "the collective talents, skills, experience, and willingness to work" (p. 209)[15] and includes intellect, intuition, productivity, creativity, cooperation, sharing, empathy, compassion, kindness, and love. Such capital is essential to the creation of healthy businesses and jobs; nonprofit organizations that educate, promote valuable causes, and provide essential services; the conscious use of resources; and the maintenance of civility and order.

Human capital also includes social networks. Many believe that community resiliency and safety depend primarily on social networks or how well connected residents are with one another. Neighborhood identity, block clubs, neighborhood watch programs, community gardens, and well-used public meeting places are all indicators of a kind of social infrastructure that lies at the heart of community resiliency—a sort of true "homeland security." This notion is dramatically illustrated by the story that follows.

In July 1995, Chicago experienced a severe heat wave. The death toll reached 739—seven times as many as died in "superstorm Sandy" in 2012. Researchers found that seven of the ten communities with the highest death rate were virtually all African American, with high rates of poverty and violent crime, while three of the ten neighborhoods with the lowest heat-wave-related deaths were characterized in virtually the same way. Englewood had 33 deaths per 100,000 residents, and neighboring Auburn Gresham's death rate was only 3 per 100,000, making it safer than many of the most affluent neighborhoods on Chicago's North Side.

The difference between these two communities was that Auburn Gresham had sidewalks, shops, restaurants, community gardens, and "community organizations that [brought] people into contact with friends and neighbors." According to one resident: "During the heat wave, we were doing wellness checks, asking neighbors to knock on each other's doors." People knew their neighbors, something that is now commonly absent in many modern American cities. People participated in block clubs and church groups. Englewood's residents, on the other hand, were vulnerable, not only because they were black and poor but also because many people and businesses had left the community, and it appeared as if the place had generally been abandoned, leaving many of the elderly vulnerable and thus afraid to leave their homes. (page 35)[16]

Reinforcing or building social infrastructure, so clearly absent from Englewood, is a fundamental goal of this element of the comprehensive plan. Since a municipality's principal charge is to protect the health, safety, and well-being of its residents, this element examines housing, public facilities and utilities, public services like fire and police, food and nutrition, and arts and culture.

For each of these components (and there could be more), we provide one or more examples of a related innovation accomplished through local public initiative. We have sought examples that we believe align with the principles, concepts, and values of a local living economy and, to repeat, rest on three principles derived from nature: zero waste, maintenance or enhancement of diversity, and the flow of "capital," as self-reinforcing feedback loops, where each is in service to the others.

Housing

It is incumbent on a city to ensure that a housing supply is diverse, including an adequate portion that is affordable, and that quality construction contributes to the longevity of the housing stock.

Ensuring broad access to housing means ensuring an adequate supply of affordable housing. The U.S. Department of Housing and Urban Development defines affordable housing as "housing that costs no more than 30 percent of a household's monthly income. That means rent and utilities in an apartment or the monthly mortgage payment and housing expenses for a homeowner should be less than 30 percent of a household's monthly income to be considered affordable."[17]

Requiring that a certain segment of the housing supply remain affordable can be accomplished through administrative fiat or law, as illustrated by the state of California's Affordable Housing Law. It can also be created through incentives. While affordability is an important commitment on the part of a community, sustainability remains the common denominator in all municipal policies and laws in a local living economy. Thus, aiming toward a supply of affordable housing that is both green built or green retrofitted would then fulfill the criteria of zero waste, diversity, and capital/energy flow.

To illustrate, the city of Bellingham, Washington, Housing Authority contributes housing to the city's supply that is both affordable *and* green, in part because grants and loans are available through the Department of Housing and Urban Development's Office of Affordable Housing Preservation. They are available to eligible property owners wishing to make efficient energy and other green retrofit investments in their property. As a condition of the loan, the property must be maintained to ensure the continued operation and maintenance of energy efficiency.[18]

The cost of energy in Bellingham was reduced by 30 percent in 1970s-era dwellings with the installation of photovoltaic solar panels, new boilers, and other measures as a result of the city's Housing Authority taking advantage of the Department of Housing and Urban Development's Green Retrofit program. These updates made buildings 17 percent more efficient than newer dwellings built to 2004 energy codes.

The Department of Housing and Urban Development's Green Retrofit Program upgraded almost 20,000 affordable homes in 200 properties, with an average energy savings reduction of 27 percent. These cost-saving measures help to preserve and ensure housing affordability.[19]

Ensuring housing quality in a local living economy means setting construction standards that produce housing that is energy efficient, durable, free from toxins, and built with materials and processes that protect the environment—or even give back to the environment. This is an essential part of a sustainable built environment.

Progress toward a sustainable built environment has been served by an ever-evolving array of building certifications. Leadership in Energy and Environmental Design (LEED) consists of a suite of rating systems for the design, construction, and operation of high-performance green buildings, homes, and neighborhoods. The U.S. Green Building Council developed LEED in the mid-1990s.[20]

Leadership in Energy and Environmental Design certification and other "green" building practices involve processes that are "environmentally responsible and resource-efficient throughout a building's life-cycle: from the setting to the design, construction, operation, maintenance, renovation, and demolition."[21] This requires close cooperation of the design team, architects, engineers, and the client at all stages of the project. Green building practices expand and complement the building design with respect to its economy, utility, durability, and comfort.

Setting the bar higher in this arena is the Living Building Challenge.[22] This building certification process is a philosophy, advocacy tool, and certification program that addresses development at all scales. It is comprised of seven performance areas: site, water, energy, health, materials, equity, and beauty. These are subdivided into a total of 20 "imperatives," each of which focuses on a specific sphere of influence. These include net-zero energy, car-free living, urban agriculture, democracy and social justice, limits to growth, and human scale. It was launched by the Cascadia Green Building Council and created by Jason F. McLennan and Bob Berkebile. It has subsequently become a part of the International Living Institute, whose mission is to provide "a visionary path to a restorative future." A *living building* is one that

- Integrates and emulates natural processes
- Obtains all necessary resources for operation from the natural environment (rainwater, wind, sunlight)
- Achieves a net-zero impact on the environment
- Produces its own energy from renewable sources
- Releases zero greenhouse gas emissions
- Functions with the water that falls on the site, capturing rainwater for interior and exterior uses

- Treats all wastewater on site
- Creates zero net waste and sources sustainable materials and thus can function autonomously from power grids and municipal water systems[23]

Living Building Challenge certification is growing in popularity but faces challenges associated with local building codes, perceived costs, and simply being a new idea. Like other local living economy initiatives, making inroads into existing planning and development policies and laws will be required.

Nevertheless, some places are embracing the concept, examples of which are explored in a further chapter. Currently, however, Washington University's Tyson Learning Center in St. Louis, Missouri,[24] and the Bullitt Center in downtown Seattle, Washington, offer examples of buildings certified to meet the Living Building Challenge.

Public Utilities and Facilities

In a municipality, public capital encompasses the aggregate body of government-owned assets that are used to promote the health, safety, and well-being of people, of commerce, and, in a local living economy, of the natural environment. Public capital includes streets and highways, airports, water treatment facilities, telecommunications, energy utilities, municipal buildings, public hospitals and schools, police, fire protection, courts, and more.

The Institute for Sustainable Infrastructure developed a checklist for evaluating the sustainability of all infrastructure projects. It represents a rating system and provides a checklist organized into 5 categories and 14 subcategories. This framework assesses the impact of a project in five areas: quality of life, leadership, resource allocation, natural world, and climate and risk.[25] Each of these categories is consistent with an ecosystem approach to planning and development, aligning with the principles of zero waste, connectivity, and maintenance of biodiversity.

Wastewater Treatment

The technology and knowledge for achieving improved water quality while creating or enhancing wildlife habitat has been available for at least a decade. In a 1993 report, *Constructed Wetlands for Wastewater Treatment and Wildlife Habitats*, 17 projects are described. They include systems involving both constructed and natural wetlands and creation and repair of habitat that lead to the improvement of municipal effluent, urban storm water, and river water quality. Although these projects are land intensive, their cost is competitive with the development of conventional treatment facilities.

The city of Cannon Beach, Oregon, for example, faced the need to replace or upgrade a three-cell sewer lagoon complex that no longer met the more stringent effluent quality standards set by the Oregon Department of Environmental Quality. After 8 years of a highly participatory process aimed at finding a cost-effective, ecologically interactive method for treating wastewater, the U.S. Environmental Protection Agency funded a treatment system that uses an existing wooded wetland to provide the final stage of the treatment process. This project was funded through the Environmental Protection Agency's Innovative/Alternative program.[26]

What is notable about this success story is that the city committed to pursuing a biophysical solution instead of a more high-tech treatment solution to a failing system, thus responding to the requirements of residents while honoring natural habitat.

Water Supply

The United States has enjoyed universal access to abundant supplies of potable water since the middle of the twentieth century. Recent, serious, and sustained droughts in the Southeast and Southwest, however, serve as a warning that this luxury may not be something we can continue to expect. Population growth, in addition to climate change, which is predicted to decrease water supplies due to decreased snowpacks and drier regional climate patterns, are facts to consider when planning for continued availability of potable water. The United States has been identified as a country that faces imminent water shortages. A Government Accountability Office survey found that water managers in 36 states anticipate shortages of water during the first two decades of this century.[27]

Like all of the examples of local government initiatives that we believe would support a local living economy, providing an adequate supply of water for human consumption, as well as irrigation and other necessities, requires a sustainable approach that serves humans, the environment, and the local financial capacity. Collecting and storing rainwater, for example, has significant potential for meeting human requirements simultaneously associated with cost savings and environmental benefits.

Our current water use practices include the use of potable water for virtually every end use, even when lesser-quality water would be sufficient. Thus, in addition to concentrated measures to increase water conservation, municipalities need to consider alternative methods for increasing the supply. Collecting rainwater from impervious surfaces and storing it for later use is a technique that has been around for hundreds of years. The use of rainwater thus collected offers a number of benefits: (1) provides an inexpensive supply of water; (2) augments the supply of drinking water; (3) reduces storm water runoff and pollution; (4) reduces erosion in urban environments; (5) provides water that needs little treatment for irrigation or nonpotable indoor uses; and (6) helps reduce peak summer demands for water.[28]

For the collection and storage of rainwater to grow as an accepted practice, specific codes need to be developed, similar to those for the use of reclaimed gray water. These would establish acceptable uses for rainwater and corresponding treatment requirements. Understanding the relationship between collecting rainwater, reduced water demand, and energy savings may catalyze the development of codes to regulate the use of such water.

> The connection between water and energy is often overlooked, but the process of extracting water from surface or groundwater supplies, bringing it to treatment facilities, treating it to drinking water standards, and delivering it to residential and commercial customers expends energy primarily because of pumping and treatment costs. The water sector consumes 3 percent of the electricity generated in the U.S. and electricity accounts for approximately one-third of utilities' operating costs. Reducing potable water demand by 10 percent could save approximately 300 billion kilowatt hours of energy each year. (p. 8)[29]

In addition, the reduction of carbon dioxide emissions associated with the use of rainwater, while not on an order of magnitude required to have a significant impact on climate change, does demonstrate the connection between the use of potable water and the required amount of energy in the broader context of sustainable water management.

The King Street Center in Seattle is a prime example of a public initiative in this realm. It uses rainwater for flushing toilets and for irrigation. Rainwater from the building's roof is collected in three 5,400-gallon cisterns. The collected rainwater then passes through each cistern and is filtered prior to being pumped through separate pipes into the building's toilets or the irrigation system. Working on an as-needed basis, the system substitutes potable water to make up the difference when the rainwater becomes low in the cisterns. The collection and reuse system provides 60 percent of the annual water needed to flush the toilets, thus conserving approximately 1.4 million gallons (5,292,000 liters) of potable water annually.[30]

Energy

The most sustainable power-generating facilities produce electricity as well as energy for other purposes, such as the production of fertilizers, building materials, livestock feed, and so on. The story in Chapter 3 about Kalundborg, Denmark, attests to human ingenuity while providing an illustration of such a multiple-purpose power-generating facility.

Reports of successes, as well as failures, related to innovations in alternative energy production are now on the rise because the acknowledgment of peak oil has spurred creativity and because of the availability of funds for experimentation, tests, demonstration projects, and so on.

Technologies that are advancing the capture and distribution of energy from the wind, sun, offshore wind, wave, geothermal, and even the transformation of sewage, plastics, used tires, as well as biomass from forestry, agricultural, and the production of livestock may well evolve in the coming decades to assist in creating a zero-waste century with low-cost energy.

Le Sueur County, one of 11 Minnesota communities comprising the Minnesota Municipal Power Agency, provides one such example. The power agency, which provides its members with "competitively priced, reliable, and sustainable energy"[31] chose Ottawa Township in Le Sueur County as the site for a biomass power-generating facility. Operating since 2012, this utility-scale energy production facility uses agricultural processing biomass to make renewable energy. The Hometown BioEnergy facility has the capacity to produce 8 megawatts of renewable electricity and is intended to help the Minnesota Municipal Power Agency meet the state-mandated renewable energy requirements of 12 percent by 2012, increasing to 25 percent by 2025. The output of the facility includes three products:

1. Electricity: Biogas will be used to produce electricity, which will be delivered directly to Le Sueur's electric distribution system.
2. Liquid Fertilizer: Nutrient-rich liquid fertilizer coproduct will be sold to local farmers.
3. Solid Fuel: Undigested biomass will be dried to create solid fuel, which will be burned off site in biomass and coal-fired boilers to replace coal.[32]

Energy conservation is another arena in which local municipalities can formulate and lead initiatives. Davis California's Cool Davis, Reducing Davis's Carbon Footprint,[33] is one such initiative. In 1999, Davis joined a small group of cities calling for local action and a national policy on climate change. In 2006, the city joined the U.S. Conference of Mayors Climate Protection Agreement that called for local and national action to reduce greenhouse gas emissions. In spring 2007, the Davis City Council unanimously adopted a strategy to reduce the city's greenhouse gas emissions.

Subsequently, in 2008, the city passed a resolution based on the findings given next to substantiate the adoption of greenhouse gas emissions targets for the Davis community. *Note*: We have purposefully included the resolution in its entirety as a model for other communities interested in making reductions not only in their use of energy but also in their contributions of carbon dioxide to the atmosphere:

> WHEREAS, there are significant long-term risks to the economy and the environment of the United States, California, and the City of Davis from the temperature increases and climatic disruptions that are projected to result from increased greenhouse gas concentrations; and
>
> WHEREAS, the potential impacts of global climate change, including long-term drought, famine, mass migration, and abrupt climatic shifts, may lead to international tensions and instability in regions affected and thereby have implications for the national security interests of the United States as well as security, economic, and environmental interests of the State of California and the City of Davis; and

> WHEREAS, local governments greatly influence their community's energy usage by exercising key powers over land use, transportation, building construction, waste management, and in many cases energy supply and management; and
>
> WHEREAS, local government actions taken to reduce greenhouse gas emissions and increase energy efficiency provide multiple local benefits by decreasing air pollution, creating jobs, reducing energy expenditures, and saving money for City government, for its businesses, and for its citizens;
>
> NOW, THEREFORE, BE IT RESOLVED, the City Council of the City of Davis does hereby adopt the following greenhouse gas emissions targets for the Davis community.[34]

A primary goal of the Cool Davis strategy is to engage 75 percent of Davis's households in reducing their use of energy through a variety of means, including the Low Carbon Diet, installing solar, retrofit ramp-ups, eating local food, biking, and more.

The carbon diet plan involves gathering 100 households into small teams to lose 5,000 pounds (2,268 kilograms) of carbon dioxide emissions in 30 days. Over 100 households answered the challenge, with participation ranging from average Davis households to city council members, city staff, University of California administrators, scientists, students on the Davis Campus, and local business owners.

The program targeted community groups and clubs, falling into three categories: education/schools/parent-teacher associations (PTAs), faith-based groups, and civic/environmental groups. On conclusion of the program, households reported the amount of carbon emissions they were able to lose through an anonymous online reporting tool. Of the 47 survey responses received, 253,723 pounds (115,087 kilograms) of carbon have been saved annually as a result of this program. This calculates to an average of 5,398 pounds (2,449 kilograms) saved per household, which exceeded the goal of 5,000 pounds (2,268 kilograms).

The city measured local greenhouse emissions using the International Council for Local Environmental Initiatives Clean Air and Climate Protection software. Based on staff research, this initiative's model is the standard used by communities to measure local greenhouse gas emissions. It was the best tool available at this time. The city's inventory is considered a work in progress, however, and will be improved when better data or methodology are available with which to significantly improve the reliability of the results.[35]

Another example of energy-related public initiative is the municipal utility of Moorhead, Minnesota, which runs a customer option called Capture the Wind. It allows Moorhead customers to choose wind power. Ten years ago, when the program launched, the wind power option sold out in less than 2 weeks and financed the construction of a 750-kilowatt wind turbine.[36]

Waste Management

Although the stated goal of a municipality may be to manage waste, there is no such thing as "waste" in nature. Therefore, what municipalities are really doing is transforming the concept of waste into "potential energy" or "fuel for other living systems." By doing so, local government is taking an action by modeling after nature (= recycling), increasing the supply of production materials, redirecting unwanted materials from landfills, and thus protecting the environment.

Whereas recycling programs are now a part of most U.S. municipal waste programs and are growing in number, incentives to individuals and businesses to participate are critical, as is happening in Westland, Michigan, where the city's curbside recycling program took a leap forward when partnered with Recyclebank and their state-of-the-art incentive. Every household's blue recycling bin has a chip on the front that registers each home on pickup day as a home that recycled. Because each household is rewarded based on the number of households that recycle and the collective amount that is recycled each weekly pickup, people are encouraged to ask friends and neighbors to recycle. They can register on the Recyclebank website and can reap rewards for different levels of points obtained through their efforts.[37]

Jonathan Hsu, chief executive officer (CEO) of Recyclebank, says of the organization:

> We know people across the country really care about their neighborhoods and the environment, and want to do their part to live a little greener. But, they just don't know where to start. That's where we come in. At Recyclebank, we help you learn how to enjoy a more sustainable lifestyle and even get rewarded for it! We bring together communities, waste haulers, and local and national brands to help us realize a world where nothing is wasted.[38]

San Francisco's recycling law, however, leads the way in being perhaps the most aggressive in the nation. In 2009, the County Board of Supervisors voted 9–2 to approve Mayor Gavin Newsom's proposal for the most comprehensive mandatory composting and recycling law in the country. An aggressive push to cut greenhouse gas emissions, its goal is was to have the city sending nothing to landfills or incinerators by 2020. The legislation requires every residence and business to have three separate color-coded bins for waste: blue for recycling, green for compost, and black for trash. Failing to properly sort refuse would result in a fine after several warnings.[39]

In January 2013, San Francisco Mayor Ed Lee reported that the city's recycling and composting law has helped the city keep 80 percent of its waste out of landfills, noting that the national recycling average is just 35 percent.[40] But, Lee wants the city to go even further. He wants to achieve zero waste. The overall strategy includes significant outreach in the community to ensure that the tough recycling laws are being followed. "Teams of workers from the city are knocking on doors of residents who, unbeknownst to them, have had their garbage cans inspected by auditors early in the morning" and are being informed that they had put items in the wrong bins.

In addition to the energy and cost savings associated with this program, jobs are being created. San Francisco's 80-year-old private garbage company recently renamed itself Recology and is investing in recycling and composting facilities, including a new, sprawling 22-acre (9-hectare) composting facility northeast of San Francisco. Approximately 400 tons (363 metric tons) of food scraps and yard clippings are brought daily to the facility and are turned to rich compost used by vineyards in Napa and Sonoma. According to Recology's CEO Mike Sangiacomo, "In terms of food waste composting, this is as good as it gets. We're creating a product that can be used on the soil to replenish nutrients that growing food crops take out of the soil."[41] And, beyond the environmental benefits of moving toward zero waste, Recology has created 186 jobs since 2003.

Transportation

Whereas traditional planning to accommodate transportation fragments the landscape and increases greenhouse gases, to our social-environmental detriment, initiatives aimed at promoting alternative modes of transportation reduce the impact on land use (such as open space), air pollution, and traffic fatalities—both human and wildlife. Thus, alternative modes of transportation are not only beneficial to human health and well-being but also reduce the many costs associated with supporting and expanding roads and facilities related to automobile use. While many cities in the United States are conscious of the need to expand public transit systems, secure green technologies (such as hybrid vehicles for government fleets), reduce air pollution, and design and build bikeways and pedestrian ways, some cities continue to innovate and lead the way.

The city of Santa Monica, California, has long led the way with its far-sighted planning and sustainability initiatives. For example, Santa Monica's Office of Sustainability and the Environment was responsible for developing and implementing policies that integrate resource management, conservation, and sustainable practices within all city operations. In addition, it administers the Solar Santa Monica program, which seeks to achieve energy self-sufficiency through the efficient use of existing energy with the integration of solar radiation in most of Santa Monica's buildings. It also offers programs that allow more efficient use of water to residents and businesses, provides assistance to the public with all aspects of green building, and provides training to residents, students, and businesses in ways to augment sustainability.[42]

In this context, it is no surprise that the city adopted Ordinance 1604, which rewards employees for using energy-efficient modes of transportation, such as walking, carpooling, van pooling, biking, public transit, compressed work schedules, and telecommuting. Because a vast majority of vehicle use in Santa Monica is from employees commuting to and from work, Ordinance 1604 was designed to help reduce such trips. Employers in Santa Monica with more than 10 employees comply with the ordinance as follows:

Employers with 10 to 49 employees provide each employee with information about ridesharing, educating their employees about issues surrounding air quality, and alternatives to their solo daily commute.

Employers with 50 or more employees encourage them to rideshare by offering them incentives to leave their cars at home and carpool, take the bus, walk, bike, or use public transit. A yearly employee survey tracks each employer's progress in reducing drive-alone trips. (http://www.smgov.net/Departments/Transportation/transportation-management-content.aspx?id=22631).

In addition, Santa Monica is the only city in the nation to implement a mandatory Parking Cash Out Program (AB2109). This program is a state law requiring employers of 50 or more employees who lease their parking or subsidize any part of their employee parking to offer their employees the opportunity to give up their parking spaces and rideshare to work instead. In return for giving up a parking space, the employer pays the employee the cost of the parking space.[43]

Food and Nutrition

Food ranks along with shelter and clothing as a basic requirement. How can local government enhance community access to food quality and quantity? In addition to providing free space for farmers' markets and permitting community gardens on public lands, other innovations are surfacing.

Darrin Nordahl, author of *Public Produce: The New Urban Agriculture* and an urban planner, considers municipally sponsored agricultural projects a natural extension of the "postorganic/buy local movement" and believes that local governments can become an agent of change in the area of local food production. Nordahl, who works in the Community and Economic Development Department in Davenport, Iowa, poses an interesting question to city planners and residents alike: Should local food production rank right up there with planning for local housing, roads and education?[44] In answer to this question, we offer the following examples:

1. Chicago houses a 200,000-member apiary on the green roof of city hall. Sale of its honey supports local cultural events.
2. In Detroit, Michigan, 30 percent of the city's land is vacant. Community groups have converted these unused locations into local opportunities to produce food.

3. Seattle has adopted a citywide goal of creating a dedicated garden site for 2,500 households.
4. Providence, Rhode Island, has the goal of doubling the amount of food grown in and around the city in the next 10 years.
5. Des Moines, Iowa, has already moved beyond public food gardens to establish public orchards, grape arbors, and a nuttery.[45]

Nordahl argues that "the sheer abundance of land within public control necessitates a hard look at how it can best serve the needs of the shareholders" and points to an "increasing number of public officials across the country who believe growing food is not only an acceptable land use, but necessary for the health and well-being of the community." In terms of land use, Nordahl contends that the future will be "a time when growing food constitutes the highest and best use for land."[46]

While most Americans enjoy daily access to food, we need to be mindful that this is a luxury, in a global sense, and like other things taken for granted, may not always be a "given." With approximately half of food bought at supermarkets currently going to landfills in the United States, according to Jared Blumenfeld of the U.S. Environmental Protection Agency, food is currently an abundant resource for many in the United States.[47] Designing and creating backup systems for necessities, however, is part of wise stewardship of a local living economy.

Cultural Resources

> There is no better indicator of the spiritual health of our city, its neighborhoods, and the larger region than the state of the arts. The arts deepen our understanding of the human spirit, extend our capacity to comprehend the lives of others, allow us to imagine a more just and humane world. Through their diversity of feeling, their variety of form, their multiplicity of inspiration, the arts make our culture richer and more reflective.
>
> **—Jonathan Fanton, President, MacArthur Foundation[48]**

Arguably, maintaining a community's culture may be the single most important thing a municipality can do to retain resiliency in the face of natural disasters, economic downturns, and other socially leveling events. The careful tending of a community's culture strengthens the social infrastructure of "place." While social media have revolutionized virtual connections between people, face-to-face contact is irreplaceable and can be facilitated through such things as government-sponsored events honoring history, culture, and the arts. In addition, government-owned and operated facilities, like community centers, recreational centers, parks, and playgrounds are all part of the municipally created commons, which by definition become part of the cultural commons and thus venues for social infrastructure to be built, strengthened, and nurtured.

A municipal cultural plan is different from, but once established dovetails with, a community cultural plan, which is developed, led, and "owned" by the community, with the municipality simply one of many stakeholders. A municipal cultural plan is the product of a municipally led process with the goal of establishing culture as an important part of economic development and community enrichment. It touches on the work of all city departments and seeks to integrate culture into formal policies, plans, and investments. For the process to be truly successful over the long term, it must engage broad representation and diversity of planning "partners" throughout the community.

Every place must define its own culture, including the identification of cultural values. As such, the cultural characteristics of place include unique social, physical, and historic qualities. Culture includes local customs, the arts, sports, recreation, chosen lifestyles, common values, and the "sense of place" with which residents and visitors alike identify. It is typically what people remember most about a place since it resides more in the heart than in the head. It is how we feel about being "there." It is what is missing when Gertrude Stein writes about there being "no there there."[49] To help illustrate cultural resource planning, we cite three prominent examples: (1) Peterborough, Ontario, Canada; (2) Seattle; and (3) Austin, Texas.

Peterborough, Ontario, Canada

(1) Peterborough, Ontario, Canada's municipal cultural plan recognizes the arts and culture as a primary driver of resiliency in both the economy and community in general. "While cultural planning has traditionally focused on the provision of facilities and programs (museums, galleries, theatres, public art, and festivals), today's cultural planning is far more place-based and recognizes that culture, the economy, and social well-being are fundamentally entwined" (p. 19).[50] Moreover, many places around the world have witnessed the positive correlation between investment in culture and economic prosperity, which means that investing in culture makes good business sense.

In a local living economy, investment in intellectual creativity in the form of ideas, innovations, and knowledge replaces capital investment in large plants and heavy machinery as the primary drivers of economic growth. "In this new 'creative' or 'knowledge' economy, as it has come to be known, the nature of work is increasingly shifting from repetitive, routine-oriented work, to work that requires creative thinking, and analytical skills, the ability to problem solve, innovate, and collaborate."[51]

Peterborough's cultural plan acknowledges the influence and work of leading economist Richard Florida, director of the Martin Prosperity Institute in Toronto. Florida has identified the people who work in the knowledge industry as the "creative class." He argues that in the new economy, the traditional economic growth model that

> courts large-scale business investment in order to attract jobs and people is no longer relevant. Instead, it is increasingly recognized that the clustering of talented and creative people in specific locales is what drives economic growth and development in those places. Where people choose to live depends increasingly on the cultural amenities, experiences, and quality of place that a city has to offer (including cultural and entertainment options, and a unique natural and built heritage), rather than on the availability of employment. In other words, today, jobs follow talent, rather than the other way around. This means that, more than ever before, place matters.[52]

Seattle, Washington

(2) In July 2008, Seattle convened the Cultural Overlay District Advisory Committee in response to the "the rapid loss of arts-related spaces and activities on Capitol Hill" and the need to "devise creative ideas for long-term promotion and preservation of arts and cultural activities and spaces in Seattle neighborhoods" (p. 8),[53] A widely diverse body, members of the Cultural Overlay District Advisory Committee represented the interests of artists, art and cultural organizations, as well as neighborhoods, business and real estate owners, real estate professionals, city and county government, and finance professionals and thus represented a unique bridging of cultural needs with public and private interests.

In August 2009, the city, in recognition of the economic and cultural benefits of "effectively preserving and creating affordable opportunities for the creative workforce, for the arts and for cultural activities in Seattle," formally accepted the recommendations of the Cultural Overlay District Advisory Committee, which identified actions for the City to take in prioritizing, promoting, preserving, and creating affordable space for arts and culture in Seattle.[54] These recommendations included the following:

1. Create designated cultural districts within Seattle's neighborhoods:

   > The vision of a cultural district is a defined, limited geographic area within a neighborhood. Inside this area, developers would be able to access incentives that would allow for the creation of permanently affordable space for designated cultural uses. [The Cultural Overlay District Advisory Committee] proposes a district model that would integrate with existing planning processes, including neighborhood planning, neighborhood councils, and existing overlays, incentives and other zoning tools that are in place or in process in the neighborhoods (including affordable housing and historic preservation).

2. Develop and promote an arts and cultural space "brand" through the city's financial and land use incentives and its regulatory tools and processes.

3. Conduct outreach and build awareness about how neighborhoods can provide arts and cultural space to encourage economic sustainability, express community identity, provide community building through participation in the arts and culture, and enhance the overall quality of life.[55]

In support of these recommended actions were findings of a study in 2008 by the Urban Land Institute that named Seattle the number one real estate market in the nation, while Americans for the Arts identified Seattle as the number one city, as measured by artists and arts organizations per capita. Thus, seeking local government assistance in providing affordable spaces for artists was critical, especially in light of the city's stated desire to ensure that the arts and culture remain a vibrant component of Seattle's vitality and livability. For more than 40 years, arts and entertainment have been drivers in the region's development. In addition, according to the *Preserving and Creating Space for the Arts and Culture in Seattle* report to the city, many studies have confirmed the value and return on investment that cultural investment has given back to its community nationally (Americans for the Arts) and regionally (Arts Fund/Creative Vitality Index).[56]

Austin, Texas

Austin, Texas, was selected along with 138 other U.S. cities to participate in a study conducted by Americans for the Arts that focused solely on the economic impact of nonprofit arts and culture organizations and event-related spending by their audiences. *Arts and Economic Prosperity IV*, the most comprehensive study of its kind, was released in 2012. In its preface, Americans for the Arts president and CEO Robert Lynch wrote: "Understanding and acknowledging the incredible economic impact of the nonprofit arts and culture, we must always remember their fundamental value. They foster beauty, creativity, originality, and vitality. The arts inspire us, sooth us, provoke us, involve us, and connect us. But they also create jobs and contribute to the economy."[57]

The study concluded that nonprofit arts and culture are a $236.1 million industry in the city of Austin—leveraging $99 million in additional spending by arts and culture audiences, spending that pumps vital revenue into local restaurants, hotels, retail stores, parking garages, and other businesses.

> By demonstrating that investing in the arts and culture yields economic benefits, *Arts & Economic Prosperity IV* laid to rest a common misconception: that communities support the arts and culture at the expense of local economic development. In fact, they are investing in an industry that supports jobs, generates government revenue, and is a cornerstone of tourism. This report shows conclusively that the arts mean business! (p. 11)[58]

A follow-up report, *The Economic Impact of the Creative Sector in Austin—2012 Update*, put the results of the Americans for the Arts study into context by showing that the creative sector (as measured by employment) rose about 25 percent over the previous 5 years, a pace more rapid than the 10 percent growth for the local economy as a whole. The study results reflect the fact that

> Creativity (in its many facets) is increasingly important to the present and future regional economy. Austin is a place that appreciates creativity and culture in a variety of evolving forms, and the incredible diversity of the local scene undoubtedly serves to both attract and retain talented people. This in turn has a significant impact on business recruitment, retention, and expansion, as well as local entrepreneurship. (p. 22)[59]

Financial Capital

Comprehensive municipal plans generally contain a component that addresses economic development with monetary (financial) wealth, financial well-being, and prosperity as its focal point. Economic development takes on a new meaning when understanding our communities as living systems, where "wealth" is not just monetary worth "but the different forms of capital, that when taken together, make up the real riches" of place.[60]

Economic Development

In the context of a living system, economic development focuses on the creation of many forms of wealth, including monetary wealth through means that are socially just and are as nonextractive as possible when it comes to natural resources. Such "means" of garnering wealth are founded on zero waste, honoring and nurturing diversity, respecting and contributing to the biophysical integrity of social-environmental capital, and the natural systems on which all life depends.

Local government is in a position to nurture the structures and patterns of local economic development through policy and legislation. The successes and achievements of local nonprofit organization and economic-development initiatives generated by citizens need to be studied by local government, not only to understand the mechanisms but also to honor the desires and creativity of its citizens. Citizen-based ideas and initiatives are represented by the following: Local First Ithaca; alternative business ownership models; and fostering economic development through human-scale public improvements.

Local First Ithaca

In Ithaca, New York, for example, Local First Ithaca is a part of the national nonprofit (Business Alliance for Local Living Economies) network of local living economies.[61] The mission of the Business Alliance for Local Living

Economies is to foster local living economies. Its work redefines economic development. The work of Local First Ithaca, like that of the Business Alliance for Local Living Economies, provides a model for alternative ways of understanding, designing, and guiding local economic development.

Local First Ithaca is an inclusive network of independent, locally owned businesses and services, farmers, nonprofits, and community members. Its mission is to bring "people, businesses, and organizations together to create a successful local economy that used resources, experience, and inclusiveness to create a thriving and livable community."[62]

While some skeptics express concerns that local living economies will cut themselves off from trade outside their borders (touting this to be unsustainable), Local First Ithaca believes that local living economies not only put a priority on the local production and exchange of products and services but also recognize the need to reach beyond their borders to other communities to trade fairly in those products and services that cannot be reasonably produced at home. Respecting the unique character and culture of other places is an important value in this framework, by which, along with the exchange of goods and services, a cultural exchange and spirit of cooperation prevail.

Local First Ithaca designs actions and programs that support its definition of a local living economy as a place, to wit:

Public policies support decentralized ownership of businesses and farms, fair wages, taxes, and budget allocations, trade policies benefiting local economies, and stewardship of the natural environment.

Citizens appreciate the benefits of buying from living economy businesses and, if necessary, are willing to pay a price premium to secure those personal and community benefits.

Investors value businesses that are community stewards and as such accept a "living return" on their financial investments rather than a maximum return, recognizing the value derived from enjoying a healthy and vibrant community and sustainable global economy.

Media provide sources of news independent of corporate control, so that citizens can make informed decisions in the best interests of their communities and natural environment.

Businesses are independent, primarily locally owned, and value the needs and interests of all stakeholders while building long-term profitability. They strive to source products from businesses with similar values, with a preference for local procurement, [that] provide employees a healthy workplace with meaningful living-wage jobs, offer customers personal service and useful safe, quality products. They work with suppliers to establish a fair exchange, cooperate with other businesses in ways that balance their self-interest with their obligation to the community and future generations and use their business practices to support an inclusive and healthy

> community, and to protect our natural environment and yield a "living return" to owners and investors.[63]

Specifically, Local First Ithaca educates the public to (1) create a sustainable business; (2) vote with one's own dollars; (3) buy and sell locally; and (4) support local agriculture. These are values that local government can nurture through policy, overlay zoning districts, special districts, and so on.

Local Ithaca's "parent organization," the Business Alliance for Local Living Economies, is North America's fastest-growing network of socially responsible businesses, comprised of over 80 community networks representing 22,000 independent business members across 30 U.S. states and Canadian provinces. The Business Alliance for Local Living Economies networks create local living economies through the building blocks of independent retail, sustainable agriculture, renewable energy, green building, local zero-waste manufacturing, community capital, independent media, and local arts and culture.

The Business Alliance for Local Living Economies envisions a sustainable, global economy as a network of local living economies, building long-term economic empowerment and prosperity in communities through local business ownership, economic justice, cultural diversity, and a natural environment that functions with biophysical integrity. As such, independent businesses create wealth by engaging local people in the production, marketing, and consumption of goods, paying taxes, and reinvesting in their communities.

A publication of the Business Alliance for Local Living Economies, *Growing Local Living Economies: A Grassroots Approach to Economic Development*, is a guide to help citizens, governments, and nonprofit organizations in designing a local plan for economic development based on the guiding principles of the Business Alliance for Local Living Economies: (1) think local first; (2) increase self-reliance; (3) share prosperity; (4) build community; (5) work with nature; (6) celebrate diversity; and (7) measure what matters.[64]

This handbook, intended to empower communities, guides users from the creation of a vision; to identification of indicators, assets, and leakages; to the development of public policy.

Alternative Business Ownership Models

Marjorie Kelly, author of a groundbreaking book, *The Divine Right of Capital*, now illustrates, in *Owning the Future: The Emerging Ownership Revolution*, how a new alternative model of ownership (called "generative ownership" or private ownership, the purpose of serving the common good) can serve and build local living economies.[65]

Like the work of the Business Alliance for Local Living Economies and other local living economy initiatives, the restructuring of business, aligned with biophysical principles and the elements of sustainability, is a direction with purpose that can be fostered by local government through land use legislation, such as overlay districts, grants, and loans and through simple public recognition of the successes of "green" businesses.

Generative business models include cooperatives, employee-owned firms, community land trusts, and community banks and credit unions that "root control in the hands of people who have a natural interest in the health of their communities and local ecosystems" says David Korten in the foreword to Kelly's *Owning the Future* (p. ix).[66]

Fostering Economic Development through Human-Scale Public Improvements

In urban planner Jeff Speck's new book, *Walkable Cities*, the economic development advantage of making cities walkable and human-scale are illustrated.[67] Because most American cities develop around the need for roads and the accommodation of automobiles, downtowns may be places where driving is easy, but the same downtown may not be a place one really wants to be. According to Speck, "Making walkability happen is relatively easy and cheap; seeing exactly what needs to be done is the trick." In *Walkable Cities*, Speck provides examples of what urban planners can do, how places can and do change, and how this one characteristic always seems to be associated with economic prosperity of a place.[68]

Walkability is also a component of what Richard Florida refers to as "people climate." In the 10th anniversary edition of his pioneering book, *The Rise of the Creative Class*, Florida relates the wealth and vibrancy of place to community development that focuses on human-scale improvements. The kinds of improvement Florida writes about include such things as reclaiming disused waterfronts and industrial areas and transforming them into green spaces, thus creating a climate that welcomes and encourages people to stroll, rest, observe, and become a part of place.[69]

In the preface to his book, Florida writes:

> I was accused of confusing chickens and eggs when I said that the secret to building better, more vibrant locations was not just attracting companies with handouts and tax breaks, but building a "people climate"—and not with stadiums and generic malls, but with parks and bike paths and street-level culture that would make people's everyday lives better, improve the underlying quality of place, and signal a community that is open, energized, and diverse.

He goes on to demonstrate how forward-looking communities, both large and small, are transforming their suburbs by making them more livable by adding transit, honoring and nurturing the local arts and cultural scene, and developing pedestrian-friendly town centers.[70]

Natural Capital

This component of the comprehensive plan focuses on natural capital, the integrity of nature's biophysical systems, which forms the earth's life support system and supplies the resources humans rely on for survival. The purpose of this element of a comprehensive plan is to establish a strategy that enables us to modify the landscape as we construct housing, public infrastructure, and commercial uses while protecting ecosystem integrity and thus sustainability. The overriding goal is to maintain a productive biosphere since there is no substitute. With respect to "natural capital," Jay Moynihan writes, "Natural capital is a term for the wide range of materials, processes and functions that underlie human society that we rely on but do not make," and we add, *cannot* make.[71]

So far, we have presented the broader context for land use in a community. Knowing how land use relates to all three "nodes" of community as a living system—natural, human, and financial—is essential in framing a sustainable approach to land use. We thus dedicate a separate chapter to natural capital. The following chapter focuses on biodiversity or the heart of ecosystem integrity—the reservoir of natural capital—as it relates to what a community does to protect soil, air, water, and the biophysical composition, structure, and function of landscapes, including natural habitat for all generations.

Summary

Chapter 4 is a critical appraisal of what it takes to plan for a local living economy by reinventing the comprehensive plan in terms of human-community requirements. Chapter 5 examines the biophysical components of nature that are required to create a community as a local living economy and thus ensure social-environmental sustainability as much as humanly possible.

Endnotes

1. Vandana Shiva. *Earth Democracy: Justice, Sustainability and Peace*. South End Press, Cambridge, MA, 2005.
2. Ibid.
3. David Korten. *Agenda for a New Economy: From Phantom Wealth to Real Wealth*. Berrett-Koehler Press, San Francisco, 2009.
4. Shiva, *Earth Democracy*.
5. Paul Hawken. http://www.brainyquote.com/quotes/quotes/p/paulhawken390962.html#b5rVBeB2vkWA2fhJ.99 (accessed February 1, 2013).
6. Ecotrust: Reliable Prosperity. "What Does Reliable Prosperity Look Like?" n.d. http://www.reliableprosperity.net/ (accessed January 20, 2013).
7. "Ecosystem." http://en.wikipedia.org/wiki/Ecosystem (accessed January 6, 2013).

8. David Korten. *Agenda for a New Economy: From Phantom Wealth to Real Wealth,* 2nd edition. Berrett-Koehler, San Francisco, 2010.
9. Ibid.
10. Miami Dade College Libraries. "Max Neef's Fundamental Human Needs." October 1, 2012. http://libraryguides.mdc.edu/maxneff (accessed January 4, 2013).
11. Bill McKibbon. "Enoughness." *Resurgence and Ecologist.* n.d. http://www.resurgence.org/magazine/article658-ENOUGHNESS.html (accessed December 26, 2013).
12. City of Davis, California. "Sustainability and Open Space Conservation." n.d. http://community-development.cityofdavis.org/sustainability-and-open-space-conservation (accessed January 3, 2013)
13. "Organic." *The New Merriam-Webster Dictionary.* Merriam-Webster, Springfield, MA, 1989.
14. Chris Maser, Russell R. Beaton, and Kevin M. Smith. *Setting the Stage for Sustainability: A Citizen's Handbook.* Lewis, Boca Raton, FL, 1998.
15. Ibid.
16. The preceding two paragraphs are based on Eric Klinenberg. "Department of Urban Planning: Adaptation/How Cities be Climate-Proofed?" *New Yorker Magazine* January (2013):32–37.
17. "Affordable Housing." http://www.sandiego.gov/housing/whatis.shtml (accessed January 15, 2013).
18. Department of Housing and Urban Development, Recovery. "Green Retrofit Program for Multifamily Housing." http://portal.hud.gov/hudportal/HUD?src=/recovery/programs/green (accessed January 28, 2013).
19. HUD's Office of Sustainable Housing and Communities. "Helping Communities Realize a More Sustainable Future." 2012. http://portal.hud.gov/hudportal/documents/huddoc?id=2012OSHCAccompRpt.pdf (accessed January 28, 2013).
20. "LEED." http://en.wikipedia.org/wiki/Leadership_in_Energy_and_Environmental_Design (accessed January 28, 2013).
21. "Green Building." http://en.wikipedia.org/wiki/Green_building (accessed January 29, 2013).
22. International Living Future Institute. Living Building Challenge. May 2012. https://ilbi.org/lbc/LBC%20Documents/lbc-2.1 (accessed December 26, 2012).
23. Whole Building Design Guide. "Living, Regenerative, and Adaptive Buildings." http://www.wbdg.org/resources/livingbuildings.php (accessed December 20, 2012).
24. Tyson Center, Washington University. "Living Learning Center at Tyson Research Center." n.d. http://tyson.wustl.edu/llc/index.php (accessed December 7, 2012).
25. Institute for Sustainable Infrastructure. "Sustainable Infrastructure Checklist." (September 3, 2012). http://www.sustainableinfrastructure.org/portal/envision2/checklist.cfm (accessed December 4, 2012).
26. The preceding two paragraphs are based on U.S. Environmental Protection Agency. *Constructed Wetlands for Wastewater Treatment and Wildlife Habitat.* EPA 832-R-93-005. Environmental Protection Agency, Washington, DC, September 1993. http://water.epa.gov/type/wetlands/upload/constructed-wetlands.pdf (accessed December 12, 2012).

27. Christopher Kloss. "Managing Wet Weather with Green Infrastructure." EPA 833-F-08-010. Environmental Protection Agency, Washington, DC, December 2008. http://water.epa.gov/infrastructure/greeninfrastructure/upload/gi_munichandbook_harvesting.pdf (accessed December 27, 2012).
28. Ibid.
29. Ibid.
30. Ibid.
31. Minnesota Municipal Power. "Community Update." http://www.mmpa.org/Uploaded_Files/9a/9ac2b975-2a79-476d-ad95-717f501e1222.pdf (accessed December 27, 2012).
32. Ibid.
33. The following discussion of Cool Davis Initiative is based on City of Davis, "Sustainability and Open Space Conservation."
34. Ibid.
35. Ibid.
36. Moorhead Public Service. "10-Year Anniversary." 2009. http://www.mpsutility.com/index.php/10-year-anniversary-celebration (accessed January 7, 2013).
37. The foregoing discussion of Westland's recycling program was based on Sustainable Westland. "Curbside Recycling in Westland." 2010. http://www.cityofwestland.com/mission-green/?page_id=76 (accessed January 1, 2013).
38. Recyclebank. "Who We Are." 2013. https://www.recyclebank.com/about-us/who-we-are (accessed January 1, 2013).
39. John Coté. "S.F. OKs Toughest Recycling Law in U.S." *San Francisco Chronicle* (June 10, 2009). http://www.sfgate.com/green/article/S-F-OKs-toughest-recycling-law-in-U-S-3295664.php#ixzz2N3KUscw1 (accessed March 9, 2013).
40. The following discussion is based on Hari Sreenivasan. "San Francisco on Track to Become Zero Waste City." *PBS Newshour*. January 25, 2013. http://www.pbs.org/newshour/bb/climate-change/jan-june13/recycling_01-25.html (accessed March 9, 2012).
41. Ibid.
42. Discussion of Santa Monica's transportation initiative based on Brenton Garen. "Record 19 Finalists to Vie for Honors at 2013 Santa Monica Sustainable Quality Awards." *Santa Monica Mirror* (January 28, 2013). http://www.smmirror.com/articles/News/Record-19-Finalists-To-Vie-For-Honors-At-2013-Santa-Monica-Sustainable-Quality-Awards/36631 (accessed January 4, 2013).
43. Ibid.
44. Urban agriculture discussion based on "Green Vegetables List," Laura Genovese's review of *Public Produce: The New Urban Agriculture* by Darrin Nordahl. January 30, 2013. http://greenvegetables-list.blogspot.com/2013/01/public-produce-new-urban-agriculture.html (accessed February 3, 2013).
45. Ibid.
46. Ibid.
47. Ibid.
48. Jonathan Fanton. *Arts and Economic Prosperity III*. Americans for the Arts, Washington, DC, 2007. http://www.artsusa.org/pdf/information_services/research/services/economic_impact/aepiii/national_report.pdf (accessed February 13, 2013).
49. Gertrude Stein. *Everybody's Autobiography*. Exact Change, 2004.

50. City of Peterborough. *City of Peterborough Municipal Cultural Plan*. Arts, Culture, and Heritage Division, City of Peterborough, Ontario, Canada, 2012. http://www.peterborough.ca/Assets/City+Assets/Arts$!2c+Culture+$!26+Heritage/Peterborough+Municipal+Culture+Plan-+FINAL.pdf.pdf (accessed February 13, 2013).
51. Ibid.
52. Ibid.
53. The following discussion is based on Cultural Overlay District Advisory Committee. *Preserving and Creating Space for the Arts and Culture in Seattle.* April 30, 2009. http://www.lundconsulting.com/pdfs/Recommendations%20FINAL%20CODAC%20April%2030%2009.pdf (accessed February 5, 2013).
54. City of Seattle Legislative Information Services. "Resolution No. 31155." Space for artists. August 17, 2009. http://clerk.ci.seattle.wa.us/~scripts/nph-brs.exe?s1=&s3=31155&s2=&s4=&Sect4=AND&l=20&Sect2=THESON&Sect3=PLURON&Sect5=RESNY&Sect6=HITOFF&d=RESF&p=1&u=%2F~public%2Fresny.htm&r=1&f=G (accessed January 1, 2013).
55. Ibid.
56. The foregoing discussion is based on Cultural Overlay District Advisory Committee. *Preserving and Creating Space* (accessed January 2, 2013).
57. Economic prosperity and the arts: Austin, TX, discussion based on Americans for the Arts. *Arts and Economic Prosperity IV in the City of Austin, Texas.* Americans for the Arts, Washington, DC, 2012. http://austintexas.gov/sites/default/files/files/Redevelopment/TX_CityOfAustin_AEP4_FinalReport.pdf (accessed December 3, 2012).
58. Austintexas.gov. *The Economic Impact of the Creative Sector in Austin—2012 Update.* TXP, Austin, TX, 2012. http://austintexas.gov/sites/default/files/files/creative_sector_impact2012.pdf (accessed December 4, 2012).
59. Ibid.
60. William C. Steward. *Sierra Nevada Wealth Index.* Sierra Business Council, Truckee, CA, 1996.
61. The following discussion of Local First, Ithaca, is based on information on its website at http://localfirstithaca.org/ (accessed January 1, 2013).
62. Ibid.
63. Ibid.
64. Ibid.
65. Marjorie Kelly and William Greider. *The Divine Right of Capital: Dethroning the Corporate Aristocracy*. Berrett-Koehler, San Francisco., 2003; Marjorie Kelly. *Owning Our Future: The Emerging Ownership Revolution*. Berrett-Koehler, San Francisco, 2012.
66. Ibid.
67. The following discussion is based on Jeff Speck. *Walkable City: How Downtown Can Save America, One Step at a Time*. Macmillan, New York, 2012. http://us.macmillan.com/walkablecity/JeffSpeck (accessed January 11, 2013).
68. Jeff Speck. *Walkable City: How Downtown Can Save America, One Step at a Time*, http://www.amazon.ca/Walkable-City-Downtown-Save-America/dp/0374285810 (accessed January 11, 2013).
69. Discussion of Richard Florida's work based on Richard Florida. "Rise of the Creative Class, Revisited." June 25, 2012. http://www.theatlanticcities.com/jobs-and-economy/2012/06/rise-creative-class-revisited/2220/ (accessed January 11, 2013).

70. Richard Florida. *The Rise of the Creative Class—Revisited*, 10th anniversary edition—revised and expanded. Basic Books, New York, 2012.
71. Jay Moynihan. "What Is Capital? A Peculiarly Human Question," The Wider View. *Ashland Daily Press* (January 10, 2001).

5

Planning for a Local Living Economy: Nature's Biophysical Requirements

> To waste, to destroy our natural resources, to skin and exhaust the land instead of using it so as to increase its usefulness, will result in undermining in the days of our children the very prosperity which we ought by right to hand down to them amplified and developed.
>
> **—TheodoreRoosevelt[1]**

Protecting Nature's Free Services

A local living economy requires natural capital in the form of essential ecosystem services if a community's human necessities are to be met in a sustainable manner. Knowledge of these services will help frame land use planning and thus help to insure that these services are kept intact to serve the community through the generations. These services are both free and essential, a rare combination of traits. It is for this reason, perhaps, that they are often overlooked in processes of valuation—monetary and otherwise.

The Millennium Ecosystem Assessment report 2005 distinguishes four categories of ecosystem services: (1) supporting services or ecosystem services "that are necessary for the production of all other ecosystem services," like nutrient cycling, soil formation, and pollination; (2) provisioning services or "products obtained from ecosystems," such as food, water, wood, fiber, fuel, and medicines; (3) regulating services or "benefits obtained from the regulation of ecosystem processes," such as waste absorption; pest control; climate regulation; flood regulation; and the capture, infiltration, and storage of water; and (4) cultural services or "nonmaterial benefits people obtain from ecosystems through spiritual enrichment, cognitive development, reflection, recreation and aesthetic experiences."[2]

Sound economic development in this context means sustaining these services while making conscious use of them with full recognition that they are not only irreplaceable but of intergenerational importance.

Protecting Diversity through Constraints to Development and Land Use Planning

Central to the integrity and sustainability of natural capital is biodiversity in all its aspects. While is not within the scope of this book to deal at length with the myriad ways a community or society can protect the diversity of nature—the wealth of society—for the benefit of all generations, we shall do our best to point out an array of options that we think will bear fruit of sufficient quality to make their cultivation worthwhile.

Habitat

The protection of habitat as part of our land use practices is used as the example because quality habitat is the basis of biological, genetic, and functional diversity, the sum of which is the basis of natural wealth and thus economic viability, which in turn equates to long-term community well-being. Habitat is composed of food, water, shelter, space, and privacy. The quality of the habitat depends on the quality and interconnectivity of these five items. Further, the environment dictates the composition of the species of plants, which creates a particular structure, in turn allowing processes and functions to occur within time and space to create the living portion of habitats for wildlife and people.

People and nature are continually changing this ecosystem or that ecosystem by manipulating the composition of its plants and thereby the system's structure and function, which subsequently changes the composition of the animals dependent on the structure and function of the resultant habitat. By altering the composition of plants within an ecosystem, people and nature alter its structure, which in turn affects how it functions, in turn determining not only what kinds of and how many animals can live there but also what uses humans can make out of the ecosystem. Understanding and maintaining the viability of these components is therefore key to the sustainability of quality habitat—including ours.

If we are really serious about achieving sustainability within our communities, we must learn to understand, accept, and act on the notion that whether populations of indigenous plants and animals survive in a particular landscape depends on the rate of local extinctions from a patch of habitat and the rate with which an organism can move among existing patches. Those species living in habitats isolated as a result of fragmentation, from such things as urban sprawl, are therefore less likely to persist. Fragmentation of habitat, the most serious threat to biological, genetic, and functional diversity, is the primary cause of not only the often-discussed global crisis in the rate of biological extinctions but also the less-discussed crisis in the rate of local extinctions.

Therefore, modifying the connectivity among patches of habitat strongly influences the abundance of species and their patterns of movement. The size, shape, and diversity of patches of habitat also influence the patterns of species abundance, and the shape of a patch may determine what species can use it as habitat. The interaction among the processes of a species' dispersal and the pattern of a landscape determines the temporal dynamics of the species' populations. Local populations of organisms that can disperse great distances may not be as strongly affected by the spatial arrangement of patches of habitat as are species that are more sedentary.

The current trend toward homogenizing the landscape, which may help maximize short-term monetary profits for one generation, progressively degrades the long-term biological adaptability of the land and thus the long-term sustainability of society. To create viable, culturally oriented landscapes, we must shift our focus from fragmentation to *connectivity* of habitats in the form of well-planned and protected systems of open spaces if we are to have adaptable landscapes with a desirable quality of living, including our indigenous wild neighbors, to pass forward to our heirs.[3]

Open Space

Open space, like water, is available in a fixed amount and therefore needs to be considered a built-in constraint. Unlike water, however, open space is visibly disappearing at an exponential rate. Once gone, it is gone, unless, of course, rural communities, and perhaps even cities, are torn down to reclaim it—an unlikely event.[4]

Space was once sacred to indigenous peoples, but today it all seems to have a price and to be coveted for that price. Whether it is "outer space, inner space, sacred space, forbidden space, your space, or my space, the more removed we are from original participation with space [which included the sanctity of all space], the more all space will continue to be desecrated," writes poet Geoffrey Hill.[5]

Consider, for example, the forested areas around Puget Sound in western Washington State, where the forests have been thinned so dramatically in recent years that "this land of towering evergreens is now relatively treeless."[6] Researchers from American Forests, one of the nation's oldest conservation organizations and based in Washington, D.C., found, through the use of satellite imagery, that nearly one-third of the most heavily forested land around Puget Sound has disappeared since the early 1970s.

Satellite photographs from 1972, 1986, and 1996 and computer-mapping software were used to study a 700-square-mile (1,813-square-kiolmeter) area that stretched across King, Pierce, Snohomish, Thurston, and Kitsap counties. Overall, the heavily forested areas, those where trees covered more than half of the land, fell from 49 percent of the region to 31 percent, a loss of about 600,000 acres (242,811 hectares). According to the study, places where trees covered 20 percent of the landscape or less increased simultaneously from 25 percent of the study area to 57 percent, an increase of more than 1 million acres (404,686 hectares). Why? The answer is simple: Subdivisions, driven by growth in the human population around Seattle and other suburban cities, have gobbled up the available open-space land.

"If people want to know why we are having so many more landslides, if people want to know why it seems to be getting hotter and why rainstorms are more intense, well, this [deforestation] is part of the answer," according to Clement Hamilton, director of the Center for Urban Horticulture at the University of Washington in Seattle.[7] But, how do fewer trees make it rain harder?

Trees provide shade, which lowers the temperature of the air. As areas are deforested, they create a phenomenon called "heat islands," where temperatures can increase 5 to 10 degrees Fahrenheit (9 to 18 degrees Celsius) or more. Then, because warm air holds more moisture, heavier rains are triggered when the warm, moisture-laden air rises and cools in the atmosphere, which increases precipitation in the form of storm water.

Forested areas typically slow and absorb water from storms, allowing it to infiltrate deep into the soil instead of flowing overland or gushing into gutters, storm drains, and water treatment plants. According to the study, it would cost about $2.4 billion to build a storm water system that would be equivalent to the one provided inherently by the trees lost since 1972. In addition, those trees would have annually absorbed, free of charge, 35 million pounds (15,875,732.95 kilograms) of pollutants, such as carbon monoxide, ozone, and sulfur dioxide. With loss of the trees, however, those pollutants, circulating freely in the air, translated into approximately $95 million in health care costs and other social impacts.[8]

Thus, while there are multiple reasons why a community might want to save open space, its irreplaceability and value added to community life are critical because in the plurality of options saved and passed forward lies the kernel of diversity and choice. Open space, as the nonnegotiable constraint around which a community chooses to develop, places the primacy of development on quality of human relationships to both people and nature. The ability and commitment to maintain a matrix of open spaces within and surrounding a community are critical to the sustainability of its quality of life (its cultural capacity, which is based on protecting its natural wealth) and ultimately the economic viability of the community, especially a small community in a nonurban setting. A well-designed, open-space system determines where both urban development and the transportation system will be located and helps to prevent a scenario such as we now see in Atlanta, Georgia.

Atlanta has become an urban heat island, experiencing temperatures up to 10 degrees Fahrenheit (18 degrees Celsius) higher than surrounding areas, creating its own weather and causing thunderstorms, according to a study by the National Aeronautics and Space Administration.[9]

Communal Open Space

Open space for communal use is not only central to the notion of community but also increasingly becoming a premium of a community's continued livability and the stability of the value of its real estate. It can also become a focal point around which to organize communities, such as the ecological restoration project in Iowa that became a community project.[10]

Here, you might wonder if an ordinary citizen can do something about saving open space. The answer is an unequivocal yes, as exemplified by Bill McDonald.[11] Bill McDonald, a fifth-generation rancher in southeastern Arizona, used his skills (in 1998) of forging consensus among fellow ranchers, some conservationists, and others, known as the Malpai Borderlands Group, to help save 800,000 acres (323,749 hectares) of open space—and with it a way of life. (For the sake of illustration, we have written this account in the present tense.)

McDonald's aim is to help ranchers become progressive trustees of an area larger than Grand Canyon National Park by keeping the connectivity of its open space intact. The parcel of land is a combination of land owned by 32 ranchers, who collectively own about half the area, and public lands the ranchers lease from the U.S. Forest Service and Bureau of Land Management, plus state trust lands belonging to Arizona and New Mexico. The ranchers' common ground—their vision, if you will—is their love of open space, the way of life it affords, and a deep desire to protect both, which means they must accept personal accountability for their own behavior. Their achievements to date (1998) are as follows:

1. Incorporating as a nonprofit organization focused on reducing polarization between the interests of ranchers and conservationists, in particular by limiting the effect of grazing livestock on public lands and in riparian zones.
2. Creating a "grassbank" and "conservation easements" intended to help ensure that the ranchers can keep their lands open and that they and their children can continue ranching without having to sell or subdivide parts of their acreages (number of hectares).
3. Emphasizing sound scientific study to discern the best approach to repairing ecologically fragile areas that will help endangered and threatened species through a "rancher's endangered species program." A working example of the program is one area ranching family trucking 1,000 gallons (3,780 liters) of water to stock ponds during a prolonged drought to keep alive one of the last populations of Chiricahua leopard frogs.
4. Getting federal officials from two states to agree to reduce fir trees in the ecosystem through prescribed burns to help bring back the area's native grasslands.

5. Improving the upland areas, including the forage on hillsides, which not only will increase the infiltration of water and thereby reduce soil erosion but also will improve the condition of riparian zones.

For his part, Bill McDonald knows that the extremists, who just want to fight, are still there. But he says, "My hope is that what we're doing will encourage other people to step toward the middle and find solutions and that the extreme positions will be marginalized over time."[12]

For such communal open space to have maximum value over time, a community, like the ranchers, must have a clear and compelling vision of what it wants so that the following questions can be answered in a responsible and accountable way: (1) Which parcels of land are wanted for the communal system of open space? (2) Why are they wanted? (3) What is their functional value: capture and storage of water, habitat for native plants and animals, local educational opportunities, recreation, aesthetics? (4) How much land is necessary to fulfill the first three questions? (5) Can one project the value added to the quality of life and the consequential value of real estate in the future? To answer these questions, and thus fulfill such a vision, requires both an understanding of the interactive differences among a vision, goals, and objectives, as well as the inviolable principles of making decisions that lead to social-environmental sustainability for all generations.[13]

Although it is not possible in this book to discuss all the possibilities of an open-space system, we consider some of the more critical and perhaps commonly overlooked components, such as the source and storage of water.

Water

Seventy-five percent of the surface of Earth is covered with water, but more than 97 percent of it is saltwater that makes up the oceans. Another 2 percent is frozen in glaciers and the polar ice caps, which means that only 1 percent of the water is available in usable form for life outside the oceans.[14]

More than 70 percent of the human body consists of water. A 1 percent deficiency of water in your body will make you thirsty, a 5 percent deficit will cause a slight fever, and an 8 percent shortage will cause your glands to stop producing saliva and your skin to turn blue; you cannot walk with a 10 percent deficiency, and you die with a 12 percent deficiency. Today, according to authorities at the United Nations, 9,500 children die every day from lack of water or, more frequently, from diseases that are carried in polluted water, so why would the U.S. Environmental Protection Agency even consider polluting water? The answer is expediency, as the example that follows illustrates.

Plans for what amounts to an underwater landfill at the mouth of the Hylebos Waterway of Tacoma, Washington, infuriated local residents.[15] To make this 1999 example more pertinent, we present it in present tense. The waterway is one of three proposed sites chosen by the U.S. Environmental

Protection Agency to dispose of a mountain of contaminated muck to be dredged over the next few years from Tacoma's Commencement Bay. The Superfund cleanup calls for extensive dredging in several of the bay's industrial waterways to remove toxic chemicals produced by the Port of Tacoma, Kaiser Aluminum, Occidental Chemical Corporation, General Metals, Elf Atochem, and Asarco, among others.

Most of the estimated 2 million cubic yards of sludge will be used as fill in two small waterways, which will subsequently be diked, capped with clean sediment, and paved to keep the toxic chemicals from leaking out, a proposal that will ostensibly improve the habitat for salmon, birds, and other wildlife. But, not so the underwater landfill proposed for Hylebos Waterway, which calls for the dumping of an estimated 300,000 cubic yards of toxic mud about 300 yards from the shore because, according to Allison Hiltner, the project manager for the U.S. Environmental Protection Agency, the area not only is big enough to hold all the mud and the underwater slope gentle enough to be stable but also the $17 million is a reasonable cost.

But, maps belonging to the state Department of Natural Resources, which owns the submerged tidal flats on behalf of all residents of the state of Washington, indicate the area may be prone to underwater mudslides, which according to Maria Peeler, manager of the aquatic resources division for the Department of Natural Resources, could break open the landfill. "It really doesn't make sense," says Peeler, "to be spending all of that money removing contaminated sediments from an area and then just shoving it somewhere else. It's like taking a broom and sweeping the dirt under a rug."[16]

Costs of the Superfund project are supposed to be paid by the industries responsible for polluting the sediments in the first place. Although the sediments could be shipped to a specialized landfill in eastern Washington, the U.S. Environmental Protection Agency rejected the idea based on logistics and the extra $7 million price tag. "They're [industry] so concerned about their costs," says Barb Bernsten, who lives along the Hylebos. "Well, I'm sorry. They're the ones who did it. Now they have to pay for it."[17]

"Why are they dumping toxic sediments in a clean area in the first place?" asks David Adams, one of some 20 families who live along the shoreline of the Hylebos in houses perched on stilts. "It just strikes us as being ridiculous."[18]

Water is a nonsubstitutable requirement of life, and its source and storage capacity are finite in any given landscape. The availability of water throughout the year will thus determine both the quality of life in a community and consequently the value of real estate. It behooves a community, therefore, to take any measure possible to maximize and stabilize both the quality and quantity of its local supply of water.

Local supply refers to water catchments in the local area under local control, as opposed to water catchments in the local area under the control of an absentee owner with no vested interest in the community's supply of water. Such absentee ownership could be a person, corporation, governing body, or agency beyond local jurisdiction.

In the western United States, for example, water pumped from deep underground aquifers is today such a valuable commodity that it is often referred to as "sandstone champagne." Much of North Africa is suffering from droughts that have forced hundreds of thousands of people to flee rural areas for low-paying jobs in cities. In Cherrapunji, a town in northern India that receives 1,000 inches (2,540 centimeters) of rain annually—the most precipitation in the world—people walk long distances to obtain drinking water, limit bathing to once a week, and have trouble irrigating their crops.

Water, not oil, will be the next resource over which nations and factions within nations will go to war. Twenty-two countries are already dependent on the flow of water that is supplied from sources in other nations. India, Pakistan, Bangladesh, the Middle East, Egypt, and Ethiopia are among the areas with the potential for armed conflict over water. There are even serious disagreements over issues concerning the sharing of water between nations on the best of terms, like Canada and the United States. In addition, China and the western United States may well have factions that are willing to compete for water in armed conflict.[19]

"It is no exaggeration to say," according to Paul Simon, former senator from Illinois, "that the conflict between humanity's growing thirst and the projected supply of usable, potable water could result in the most devastating natural disaster since history has been recorded accurately, unless something happens to stop it." That something would have to be far wiser leadership at home and abroad than we have thus far seen in our lifetimes.[20]

With this in mind, it is wise to use the storage of water for present and future generations as one of the cornerstones in any open-space system. If outright purchase of a water catchment is not possible, a community could conceivably enter into a long-term lease or contract to rent a catchment, with control over what is done on it. Then, it might be possible to accrue monthly or annual payments toward the price of purchasing the land at a later date. Such an arrangement could benefit the owner in terms of a steady income at reasonable tax rates while allowing some acceptable use of the land.

Another alternative might be a tax credit payable to the landowner if the community could work in conjunction with the owner to protect the water catchment's inherent value to the community itself. There are probably other options, such as reintroduction of treated water into a known aquifer, but the important consideration is to secure the purchase of local water catchments in community ownership as part of the open-space program to maintain and protect the quality of life and the local value of real estate. An added value may be that such water catchments in an open-space system can act as islands of quiet amidst the daily bustle of town life.[21]

Quiet

Quiet, like water and open space, can now be seen as a disappearing resource and thus another constraint. Today, there is an increasingly noticeable din from autos, jets, buses, leaf blowers, lawn mowers, and trucks that seem in perpetual motion at all hours. If a community designs its system of open spaces to dampen the constant stimulation of urban background noise, people could, for many years to come, find a peaceful quiet in which to relax and hear the songs of birds, the wind in the trees, and the roar of the ocean.

The relaxation experienced in a quiet place can be consciously enhanced in an open-space system by including farm and forestlands, riparian areas, and floodplains as buffers against city noise. I (Chris) know this is possible because I have experienced it in the beautiful Shinto shrine in downtown Tokyo, Japan, an exceedingly busy city. Likewise, Jane, who is frequently near Pioneer Square in the heart of downtown Seattle, knows the transporting effect of a vest pocket park, Waterfall Garden, featuring a 22-foot waterfall that totally shuts out the sounds of passing cars and buses.

Surrounding Landscape

The land surrounding a community's municipal limits gives the community its contextual setting, its ambiance, if you will. The wise acquisition of open spaces in the various components of the surrounding landscape, whether nature's ecosystem or culture's, protects, to some extent at least, the uniqueness of a community's setting and hence the uniqueness of the community itself. And, the value added, both spiritual and economic, will accrue as the years pass.

Some cities have natural barriers to expansion and thus development. Santa Barbara and Santa Cruz, California, two cities of extraordinary beauty, prosperity, and progressive urban planning, enjoy an eastern natural boundary represented by the Coast Range of mountains, which acts as a barrier to expansion and development.

Agricultural Cropland

A community could purchase open space in the form of fencerows along which to allow fencerow habitat to re-create itself. Then, in addition to mini-habitats in and of themselves, the few uncultivated yards along each fence could once again act as longitudinal corridors for the passage of wildlife from one area to another. Living fencerows would also make the landscape more interesting and more appealing to the human eye and again add the songs of birds and the colors of flowers and leaves to the passing seasons.

The point is to find out what worked sustainably in the past and begin re-creating it in the present, and where problems arise, as they will, to work together to resolve them. The only way to create, maintain, and pass forward the sense of community is by working together because the friendliness of a community is founded on the quality of its interpersonal relationships, of which small family farmers can be an integral part.

Forestland

If a community is in a forest setting, the forest more likely than not is a major contributor to the community's image of itself, in addition to which it may comprise an important water catchment. Furthermore, if the community is, or has been, a "timber town," then most of the forest may well have been converted into economic tree farms. Therefore, maintaining an area of native forest may be of even greater value. If some old-growth trees are included in the area, its spiritual value may well be heightened and its value as habitat for some plants and animals greatly enhanced.

On the other hand, if what surrounds a community is no longer forest but rather an economic tree farm, a purchased area could be allowed to evolve once again toward a forest. As such, its aesthetic and spiritual values will increase, as will its potential value as habitat and for educational purposes. Much can be learned by comparing a relatively sterile tree farm with a real forest.[22] One will find, for instance, that a forest harbors a far greater diversity of species of both plants and animals than does a tree farm, even one near the age of cutting.

Although we have used the forest as an example because Chris grew up in one, the same concepts can be applied anywhere. Outside Denver, Colorado, for example, is a wonderful open space that represents a vestige of native shortgrass prairie that once covered the eastern part of the state. It is beautiful. And, it is inspirational, creating, as it does, a tangible tie to a now-intangible past and an unknowable future.

Riparian Areas and Floodplains

Riparian areas and floodplains are coming under increasing pressures of urban development because of the misguided notion that we humans can unilaterally entrain streams and rivers with impunity, despite much and growing evidence to the contrary. When a levee fails, the response seems always to be more levees and, if need be, more dams. We have yet to understand that a problem caused on one level of human consciousness cannot be fixed on the same level of consciousness.

If we are willing to risk moving to a higher level of consciousness, we can either prevent or repair much of the damage our shortsighted human activities cause. Take, for example, the Snake River near Jackson, Wyoming, where engineers, people from the Jackson Hole Conservation Alliance, the Wyoming Department of Fish and Game, and local officials are planning to restore part of the Snake River to a more natural condition.[23] To better capture the essence of the story, it is related in present tense.

For nearly 40 years, levees that line about 23 miles (37 kilometers) of the Snake River near Jackson have entrained high water from the melting snows of spring in the Teton Mountains and allowed lavish homes to invade the cottonwood forest of the river's floodplain. Although researchers have known for some time of the ecological havoc wreaked by dams, only recently have they begun to recognize the ecologically destructive nature of levees and their "free-form cousin" riprap, the latter being piles of rock and earth dumped by landowners along streams and rivers to guard against erosion. It is not surprising, therefore, that the 15-foot-high (4.6-meter-high) serpentine piles of rock created by the U.S. Army Corps of Engineers to protect farmers' fields and hay meadows from flooding and erosion have caused serious and unexpected problems along one of the world's most scenic stretches of river.

The river, for its part, for centuries had been true to its rhythm of flooding and receding in a fluid motion that constantly redesigned its five or six channels or braids as it dissipated the energy of its floodwaters each spring. But now, squeezed into one or two rigid channels, the upper Snake River has lost its ability to flood during the spring runoff. This lost ability has increased the velocity of the water from spring snowmelt within the levee straitjacket, which in turn has caused the raging spring torrents to destroy many of the large islands in the remaining channels once occupied by willows and cottonwood trees. The luxuriant forests of cottonwood that lined the river's banks of yore are fading into an era of lost memory for lack of young trees to replace the dying old ones because cottonwoods need periodic flooding to reproduce successfully. And, Snake River cutthroat trout, which need clean gravel in which to spawn, have suffered from the channelization because floodwaters no longer flush and rejuvenate their spawning gravels.

In the autumn 1998, the Teton County Natural Resource District and the U.S. Army Corps of Engineers began testing methods of breathing old life back into the river by re-creating its rhythms of flooding. Such a venture would, however, take a different level of consciousness, as Rik Gay, manager of the project for Teton County, pointed out when he said, "Rivers don't just go downstream. We need to think in three dimensions. Rivers also move laterally and below the ground."

There is a sober reminder in all of this, however. The levees along the upper Snake River not only have narrowed and denuded the river over the last four decades but also have allowed million-dollar housing developments to flourish, which caused Bill MacDonald, manager of the Snake River project for the U.S. Army Corps of Engineers, to observe that returning the natural flow of the river was "not feasible" because "behind those levees are millions, if not billions, of dollars in real estate." What might an alternative for the future be?

According to Scott Faber, a floodplain expert with the conservation group American Rivers, what really needs to be done to protect the ecological integrity of rivers is to cease building in the floodplains. One way to accomplish that, he contends, is to terminate the federal subsidy for repairing levees and make local governments pay the cost.[24] In the meantime, however, many communities still have riparian areas that are important to understand and protect.

Riparian Areas

Riparian areas can be identified by the presence of vegetation that requires free or unbound water and conditions moister than normal. These areas may vary considerably in size and the complexity of their vegetative cover because of the many combinations that can be created between the source of water and the physical characteristics of the site. Such characteristics include gradient, aspect of slope, topography, soil and the type of stream bottom, quantity and quality of the water, elevation, and the kind of plant community.

Riparian areas have the following things in common: (1) They create well-defined habitats within much drier surrounding areas; (2) they make up a minor portion of the overall area; (3) they are generally more productive than the remainder of the area in terms of the biomass of plants and animals; (4) wildlife use riparian areas disproportionately more than any other type of habitat; and (5) they are a critical source of diversity within an ecosystem.[25]

There are many reasons why riparian areas are so important to wildlife, but not all can be attributed to every area. Each combination of the source of water and the attributes of the site must be considered separately. In addition, riparian areas supply organic material in the form of leaves and twigs, which become an important component of the aquatic food chain. Riparian areas also supply large woody debris in the form of fallen trees, which form a critical part of the land-water interface, the stability of banks along streams and rivers, and in-stream habitat for a complex of aquatic plants as well as aquatic invertebrate and vertebrate organisms.[26]

Setting aside riparian areas as undeveloped open space or repairing them through ecological functions[27] means saving the most diverse, and often the most heavily used, habitat for wildlife in proximity to a community. Riparian areas are also an important source of large woody debris for the stream or river whose banks they protect from erosion.[28] Furthermore, riparian areas are periodically flooded in winter, which, along with floodplains, is how a stream or river dissipates part of its energy. It is important that streams and rivers be allowed to dissipate their energy; otherwise, floodwaters would cause considerably more damage than they already do in settled areas.

Floodplains

A floodplain is a plain that borders a stream or river that is subject to flooding. Like riparian areas, floodplains are critical to maintain as open areas because, as the name implies, they frequently flood. These are areas where storm-swollen streams and rivers spread out, decentralizing the velocity of their flow by encountering friction caused by the increased surface area of their temporary bottoms, both of which dissipate much of the floodwater's energy.

It is wise to include floodplains within the matrix of open spaces for several other reasons: (1) They will inevitably flood, which puts any human development at risk, regardless of efforts to steal the floodplain from the stream or river for human use (witness the Mississippi River); (2) they are critical winter habitat for fish;[29] (3) they form important habitat in spring, summer, and autumn for a number of invertebrate and vertebrate wildlife that frequent the water's edge;[30] and (4) they can have important recreational value.

If all these kinds of areas were incorporated into a system of well-designed, well-connected open spaces, a community would be wealthy indeed. In addition, the community would have done much to maximize the quality of life (its cultural capacity), not only in the present but also for the future.

Consider, for example, that a well-implemented system of open spaces helps to ameliorate the cumulative effects of a concentrated human population on its immediate surroundings. It also ensures that some areas are protected from the intrusion of artificial structures to clutter and fragment the space, which allows the seeming "emptiness" to be filled with wildflowers, grasses, trees, butterflies, birdsong, and glimpses of wildlife in an area where they need not compete with such human endeavors as agriculture and transportation. Open space also connects people with the land and its variety of habitats and life forms. Most important, open space, as the non-negotiable constraint around which a community plans and carries out its development, allows both roads and people to be placed in the best locations from a sustainable point of view, both environmentally and socially.

Planning for future flooding in New York followed in the wake of superstorm Sandy in 2012, as engineers and planners contemplate how to prepare for future megastorms to prevent the level of damage New York continues to repair. "There are design solutions," according to architect Eric Bunge, that "transcend our current obsessive discourse of efficiency and defense."[31] In a *New Yorker* article, "Save New York by Making It 'Soft,'" writer Thomas De Monchaux notes, "That transcendent resiliency derives from changeability and permeability, and thus from a kind of dynamic equilibrium between natural and artificial elements."[32] New York's harbor, "as it exists at the present time," another engineer and landscape architect has said, embodies "the changes created by man, and by nature in her struggles with his innovations." That was Viele, writing in 1855 to the New Jersey State Legislature's Joint Committee on the Encroachments upon the Bay and Harbor of New York. He concluded, "With regard to rivers, nature acts by certain fixed laws, from which she never deviates; in accordance with these laws, their beds are established and their channels excavated. Man cannot change them; all his efforts to do so, have but produced modifications; the evil he attempts to remove, reappears with greater force at another point."[33]

The Misguided Role of Today's Planning for Transportation

When the system of transportation becomes the center of a community's vision for its future, the community is placing the primacy of its vision on maximizing the efficiency of moving masses of people from one place to another, which in turn determines where and how the population versus open spaces will be situated. Here, a fundamental question might be posed: Does building more and more roads really relieve congestion, which, after all, seems to be what drives the design of a transportation system?

According to Bill Bishop, editorial page columnist for the *Herald-Leader* in Lexington, Kentucky, building more roads does not relieve congestion—it *adds* to it.[34] We think he has a good point because there is a parallel in buying houses.

We have often heard people say they have so much stuff that they need a larger house. This statement seems reasonable enough on the surface, but in practice, most people we know who have actually bought a larger house begin immediately filling it to capacity. Why? Our American compulsion to fill every nook and cranny is in part a product of not having been taught how to live with empty space, or at least space that is not crammed full all the time.

Is the same true with roads? If our cities' roads are congested and we build more roads to relieve the congestion, will we not just fill the new roads again to the point of congestion, like our houses? It seems to us that one could logically say: Like our houses, so our roads. This is also the contention of columnist Bill Bishop.

"Trying to pave your way out of traffic congestion," writes Bishop, "is like trying to eat your way back into your high school jeans. Cars fill in the new pavement just like middle age created the market for Dockers." Although it seems counterintuitive, says Bishop, building more roads actually leads to more traffic. On the other hand, he continues, closing roads, or even narrowing streets, does not create more congestion—it tends to cut the volume of traffic, especially in cities.

"Lord knows," says Bishop, "the evidence of this phenomenon is stalled in full view of most citizens. As soon as roads are built, they're filled. And to relieve the new traffic, we build new roads. You'd think somebody would connect the dots."[35] What dots? The dots pointing to the fact that the level of consciousness that caused the problem in the first place, such as the levees along the Snake River near Jackson, Wyoming, which largely destroyed the river, is not the level of consciousness that can solve it. A higher level of consciousness is required—recognizing, accepting, and acting on the evidence under our noses, which is connecting the dots.

Some people have connected the dots, quips Bishop. "Adding new roadways and widening older ones was seen as the way to solve the problem," observed the Texas Transportation Institute in a study of city traffic. "In most cities, this new roadway capacity was quickly filled with additional traffic, and the old problems of congestion returned."[36]

On the other hand, researchers at the University College of London, England, examined 60 cases from around the world in which roads had been closed. They found that a goodly portion of the traffic that once used the roads simply "evaporated." The cars and trucks were not simply rerouted on nearby streets, but disappeared altogether.

On average, one-fifth of the vehicular use, and in some cases as much as 60 percent, went away once a road was closed, and the full volume of vehicles did not reappear once a road was reopened. The Tower Bridge in London, for example, was temporarily closed in 1994, and the traffic dispersed. Three years after the bridge was reopened, traffic still had not returned to its former level.

Writer James Howard Kunstler argues in the online magazine *Slate* that "we have transformed the human ecology of America, from sea to shining sea, into a national automobile slum."[37] "Do we get what we get," muses Bishop, "just because we can't remember any other way to live?"[38]

That is a good question because, while transportation may be the center of a community's vision for its future, which increases the artificial structures of urbanization, light and noise pollution, and simultaneously precludes much of nature through fragmentation of habitats, a community has two options in planning its transportation system: ecological constraints (greater emphasis on quality of life and thus protection of open space) or economic constraints (greater emphasis on economic efficiency with a disregard for the irretrievability of open space).

If a community chooses to plan its transportation system around existing biophysical constraints, it could have a relatively good open-space system if the focal point of the vision and its implementation is based on the ecological effectiveness of the land use design. On the other hand, if a community chooses to position its transportation system—as is currently done—to maximize its economic efficiency, a viable system of open spaces is all but foregone because habitat fragmentation increases with growth the transportation system. There is also a greater likelihood of rampant habitat fragmentation—as opposed to a community where a system of open spaces has primacy—because the transportation system would be so located to continually accommodate growth in the human population. This myopic vision would put cumulative pressure on the survival of indigenous habitats and their species. All of these things operate synergistically as cumulative effects that exhibit a lag period before fully manifesting themselves.

When a Community's Population Begins to Destroy Its Quality of Life

When a community uses growth as the primary *nonnegotiable* constraint around which development will proceed, it is most likely, in our experience, to view the size of the proposed population in terms of continual economic growth (biological carrying capacity) as opposed to social-environmental sustainability and a correspondingly good quality of life (cultural carrying capacity). Thus, a vision based on the human population as the nonnegotiable constraint usually, but not always, places its primacy on pushing the biological carrying capacity to the perceived limit, especially when money is involved.

A case in point is Corvallis, Oregon, where increased population means increased money through federal block grants, which in turn means increased growth, which again is seen as a means to garner more money. (For a discussion of the "growth/no-growth tug-of-war," see Chapter 5 in *Setting the Stage for Sustainability.*[39])

At a public forum during the late 1990s, the Business Advocacy Committee of the Corvallis Area Chamber of Commerce made public the results of its research into the management of municipal growth. According to Patricia Mulder, the executive director of the chamber, the people of Corvallis had been diligent and "perhaps" too successful in preventing rampant growth in the town's population, which had declined 3 percent in the past year with a workforce that had shrunk by 5 percent. "These declines have occurred," says Mulder, "while surrounding communities continue to grow."[40]

The crux of Mulder's concern was that the loss of 1,500 citizens left the population of Corvallis below 50,000 inhabitants and thus threatened the classification of Corvallis as an "entitlement" city following the next census. Why the concern? If Corvallis were to reach a population of 50,000, with some additional requirements, it would be "entitled" to community development block grants from the federal government—in a word, "money."[41]

There was no need to worry, however, because the next U.S. Census Bureau's estimate put the population at 50,202 people, a figure that entitled Corvallis to receive about $400,000 a year from the U.S. Department of Housing and Urban Development for housing programs, beginning the following fiscal year, to help low- and moderate-income residents. This was deemed important because the state oversees the money, and cities with less than 50,000 inhabitants must compete for it, whereas cities with a population over 50,000 are entitled to the money without competition. In other words, the city gets its own money without intercession by the state.[42]

Coupled with the Corvallis area chamber's notion of the need to "grow" the population (which, incidentally, fits precisely the notions of many other area chambers of commerce), is the perceived need for waste to keep people busy and expand the economy, so says author Barry Brooks. Here, the concept is that durability (which is a perceived economic threat to business leaders because it slows sales) equates over time to sustainability, whereas built-in obsolescence (which stimulates repeated sales and equates to waste) is good for the economy as we define it, despite the regrettable fact that planned obsolescence is counterproductive to social-environmental sustainability. This is but one more example of the misleading notion of economic well-being fostered by the flawed calculation of the gross domestic product as opposed to accounting for both credits and debits according to the genuine progress indicator.

As is often the case, the desire for more and more money drives a community's population growth—as its nonnegotiable planning constraint—to such a point that the quality of life begins an almost invisible (but irreversible) decline. Thus, a community that chooses to move ever closer to the perceived biological carrying capacity of its human population as the nonnegotiable constraint usually sees the placement of its people dictated by the desirability of available private property-based individual tastes, precluding a connected system of open spaces with any integrity of habitat. This, in turn, usually determines where the transportation system will be located, which works to maximize the fragmentation of habitat. In addition, the community maximizes the cumulative effects of light and noise pollution, artificial structures, and declines in and extinctions of local populations and species of indigenous plants and animals.

All these effects are not only hidden for some time in the ecological lag period but also work synergistically in shifting the landscape from the more natural end of the continuum to the more cultural end. Beyond some point, these human-induced, biophysical effects upset the ecological integrity and ultimately affect the quality of life, almost inevitably in the negative. So, where do we go from here?

Economics may continue to shift populations around the country, but changes in the way we use and treat the land could perhaps bring an end to this cycle. If, for example, the process of land use planning and development were one that in and of itself helped to create a sense of place within a community, then wealth and prosperity might begin to be redefined: The natural wealth of "community" might begin to meet deeper human necessities, thus diminishing the material desires that currently define the notion of *wealth*.

The coastal community of Seaside, Oregon, exemplifies an alternative to this scenario.[43] People get away from such metropolitan areas as Portland as a regular occurrence during the summer, and the Oregon coast is a place of choice to visit. That choice, however, has resulted in towns built around tourism-fueled development, which in turn brought traffic, strip malls, and sprawl, where people go from beachside hotels to downtown hot dog stands, video arcades, and shopping malls.

While other coastal towns stretch sprawl farther and farther along U.S. Highway 101 in the name of economic growth and vitality, Seaside took the opposite tack. With the idea in mind to help save salmon habitat, give local residents a place to play, and reverse a century of tourist-seeking development, a farsighted, reform-minded group in the city hall of Seaside purchased mudflats, marshes, and millponds to save what was left of nature in Seaside. Their plans even specified tearing down of a few buildings along U.S. Highway 101 to let rivers and wetlands greet visitors instead of the proverbial gift shops and sundry tourist traps.

"Originally, we thought visitors were our treasure," said Neal Maine, the resource manager for the North Coast Land Conservancy. "And then we realized that we were standing on it [meaning the land and its diversity as the community's natural wealth]."[44] "We are trying to do it for ourselves," added city manager Gene Miles. "And the truth of the matter is, it will probably be a bonanza."[45] To realize the bonanza of which Gene Miles speaks, however, one must not only consider but also account for the cultural carrying capacity of a community.

In the End, It Is a Question of Biological Carrying Capacity versus Cultural Carrying Capacity

Although we may think ourselves wise in our own eyes, we are too often blind to the truth that we neither govern nor manage nature.[46] We treat nature wisely or unwisely for good or for ill, but we do not control nature. We do something to nature, and nature responds, and in the response lies the lessons we are loathe to learn—lessons about lifestyle.

Although we may think ourselves wise in our own eyes, we are too often blind to the truth that we neither govern nor manage nature.[47] We treat Nature wisely or unwisely for good or for ill, but we do not control Nature. We do something to Nature, and Nature responds, and in the response lies the lessons we are loathe to learn—lessons about lifestyle, which pits biological carrying capacity against cultural carrying capacity. Lifestyle can be defined as an internally consistent way of life or style of living that reflects the values and attitudes of an individual or a culture. Many in Western society have made lifestyle synonymous with "standard of living," practiced as a search for ever-increasing material prosperity. If, however, we are to have a viable, sustainable environment as we know it and value it, we must reach beyond the strictly material and see lifestyle as a sense of inner wholeness and harmony derived by living in such a way that the spiritual, environmental, and material aspects of our lives are in balance with the capacity of the land to produce the necessities for that lifestyle.

Whether a given lifestyle is even possible depends on *cultural carrying capacity*, which is a notion used to describe, in *qualitative* terms, the kind of lifestyle a community wants to enjoy measured against the ability of the community's surrounding environment to sustain that lifestyle.[48] *Biological carrying capacity*, on the other hand, is strictly *quantitative* in that it represents the maximum number of animals—or people—that can live in and use a particular area without destroying its ability to function in an ecologically specific way. Where cultural carrying capacity deals with the quality of life per individual, biological carrying capacity deals strictly with maximizing the number of individuals to the threshold of habitat destruction, beyond which the population will collapse. If we want human society to survive the twenty-first century in any sort of dignified manner, we must have the humility to view our own population in terms of local, regional, national, and global biological carrying capacities because the quality of life declines in direct proportion to the degree to which the habitat is overpopulated.

If, therefore, we substitute the idea of cultural carrying capacity (quality of life) for biological carrying capacity (maximum numbers of individuals), we have a workable proposition for sustainable community. Cultural carrying capacity is a chosen quality of life that is sustainable without endangering the productive capacity of the environment. The more materially oriented the desired lifestyle of an individual or a community, for example, the more resources are needed to sustain it and the smaller the human population must be per unit area of landscape. Cultural carrying capacity, then, is a balance between the way in which we want to live, the real quality of our lifestyle and our community, and the number of people an area can support in that lifestyle on a sustainable basis.

Cultural carrying capacity (= quality) of any area will be less than its biological carrying capacity (= quantity) in the biophysical sense. Cultural carrying capacity has built into it the prudence of limitations as a margin of safety in the event of such long-term phenomena as global climate change. Biological carrying capacity, on the other hand, uses the environment to its maximum and lacks a margin of safety for difficult years or unforeseen environmental changes, which, when they occur, as they always do at some point in time, wreak havoc on the population. The long-term environmental risks hidden in the momentary notion of an area's biological carrying capacity, when exploited to the maximum, have doomed more than one civilization to collapse by destroying the biological sustainability of the surrounding landscape.

David Skrbina, a graduate student at the University of Bath, in Bath, England, states the problem a little differently.[49] The problem with most human populations, as Skrbina sees it, is one of scale. To sustain itself, a given human population must maintain a size that is both compatible with a concept of community and consistent with a human scale, which not only means a community of a size that fosters people actually knowing one another but also means a community of a size that fits with ecological comfort into its immediate landscape.

The problem of human population has two dimensions, according to Skrbina—the *size* of the population and the *density* of the population. While a large population can be a problem, "it is a *different* problem if those people are packed onto a small island than if they are spread across a large prairie."[50] Population has two factors that must be simultaneously accounted for: the number of *people* and the area of *land* available to support them sustainably to make cultural carrying capacity a workable idea.

We can, for example, predetermine local and bioregional carrying capacity in the cultural sense and adjust our population growth accordingly. (Bioregion is used as a geographically definable area of biological similarities, which is largely self-contained when it comes to a supply of water.) If we choose *not* to balance our desires with the land's biophysically sustainable capabilities, the depletion of the land will determine the quality of our cultural/community/social experience—our lifestyle. If, however, we choose to balance our desires with the land's sustainable capabilities, then we will model our planning after nature's aforementioned biophysical principles, which means humanity would leave a much lighter imprint on Earth.

Summary

Chapter 5 examines the protection of natural capital as part of land use planning that ensures social-environmental sustainability as much as humanly possible. Chapter 6 intends to unravel barriers to creativity in the planning process by discussing some possible reasons for them, which seem embedded in human nature, and how to overcome them by reframing the problem in a more positive way.

Endnotes

1. Theodore Roosevelt. http://www.goodreads.com/quotes/tag/environment (accessed February 14, 2013).
2. Millennium Ecosystem Assessment. *Ecosystems and Human Well-Being: Synthesis.* Island Press, Washington, DC, 2005. http://www.unep.org/maweb/documents/document.356.aspx.pdf (accessed February 14, 2013).
3. The foregoing discussion is based on Chris Maser. *Social-Environmental Planning: The Design Interface between Everyforest and Everycity*. CRC Press, Boca Raton, FL, 2009.
4. Ibid.
5. Geoffrey Hill. "The Sacredness of Space." *Creation Spirituality* 12 (1996):31–33.
6. The following discussion of forests in the area of Puget Sound, Washington, is based on J. Martin McComber. "Study Shows Puget Sound Forests Are Slowly Thinning." *Corvallis Gazette-Times* (Corvallis, Oregon) July 15, 1998.
7. Ibid.
8. Ibid.
9. Phil Williams and Victoria Bruce. "Atlanta, an 'Urban Heat Island' with Higher Temperatures than Surrounding Area, According to a NASA-Sponsored Study." University of Georgia Communications News Bureau, May 20, 1999.

10. Rick Patterson and Jan Aiels. "Iowa Restoration Becomes a Community Project." *Land and Water* 42 (1998):43–45.
11. The discussion of Bill McDonald is based on Arthur H. Rotstein. "Genius at Home on the Range." *Corvallis Gazette-Times* (Corvallis, OR) June 22, 1998.
12. Ibid.
13. (1) Chris Maser. *Vision and Leadership in Sustainable Development*. Lewis, Boca Raton, FL, 1998; and (2) Chris Maser. *Decision Making for a Sustainable Environment: A Systemic Approach*. CRC Press, Boca Raton, FL, 2013.
14. Unless otherwise stated, the brief discussion on the scarcity of water is drawn in part from (1) Chris's experience at home and various countries abroad; (2) Roar Bjonnes. "Sweet Water and Bitter." *Resurgence* 181 (1997):32–34; and (3) Paul Simon. *Tapped Out* [excerpt]. *Parade Magazine* August 23, 1998:4–6.
15. Associated Press. "Residents Fume over Tacoma's Underwater Hylebos Landfill Plan." *Corvallis Gazette-Times* (Corvallis, OR) December 9, 1999.
16. Ibid.
17. Ibid.
18. Ibid.
19. The preceding five paragraphs are based on Chris Maser. *Ecological Diversity in Sustainable Development: The Vital and Forgotten Dimension*. Lewis, Boca Raton, FL, 1999.
20. Simon, *Tapped Out*.
21. The foregoing two paragraphs are based on Chris Maser. *Sustainable Community Development: Principles and Concepts*. St. Lucie Press, Delray Beach, FL, 1997.
22. (1) Chris Maser. *Forest Primeval: The Natural History of an Ancient Forest*. Sierra Club Books, San Francisco, 1989; (2) Chris Maser. *Sustainable Forestry: Philosophy, Science, and Economics*. St. Lucie Press, Delray Beach, FL, 1994; and (3) Chris Maser. *Our Forest Legacy: Today's Decisions, Tomorrow's Consequences*. Maisonneuve Press, Washington, DC, 2005.
23. The following discussion about repairing part of the Snake River to a more natural condition is based on Jim Robbins. "Engineers Plan to Send a River Flowing Back to Nature." *The New York Times* May 12, 1998.
24. Ibid.
25. (1) Jack W. Thomas, Chris Maser, and Jon E. Rodiek. "Riparian Zones." In: *Wildlife Habitats in Managed Forests—The Blue Mountains of Oregon and Washington*, 40–47. USDA Forest Service, Agricultural Handbook No. 553 (Jack W. Thomas, technical editor). U.S. Government Printing Office, Washington, DC, 1979; and (2) Jack W. Thomas, Chris Maser, and Jon E. Rodiek. "Riparian Zones." In: *Wildlife Habitats in Managed Rangelands—The Great Basin of Southeastern Oregon*. USDA Forest Service General Technical Report PNW-80 (Jack W. Thomas and Chris Maser, technical editors). Pacific Northwest Forest and Range Experiment Station, Portland, OR, 1979, 18 pp.
26. Chris Maser and James R. Sedell. *From the Forest to the Sea: The Ecology of Wood in Streams, Rivers, Estuaries, and Oceans*. St. Lucie Press, Delray Beach, FL, 1994.
27. For good examples of how to repair riparian areas see (1) Steve Apfelbaum and Jack Broughton. "Applying an Ecological Systems Approach in Urban Landscapes." *Land and Water* 42 (1988):6–9; and (2) David Lee and Jim Lovell. "Urban Trout Stream Gets a Second Chance." *Land and Water* 4 (1988):16–19.
28. Geoff Mulgan. "Connexity, How to Live in a Connected World." *Resurgence* 184 (1997):6–7.

29. Maser and Sedell, *From the Forest to the Sea.*
30. Ibid.
31. Thomas De Monchaux. "Save New York by Making It 'Soft.'" *New Yorker* January 15, 2013. http://www.newyorker.com/online/blogs/culture/2013/01/how-to-protect-new-york-from-rising-waters-with-soft-infrastructure.html#ixzz2IGGXXpjg (accessed January 5, 2013)
32. Ibid.
33. Ibid.
34. The discussion of congestion and transportation is based on Bill Bishop. "To Reduce Congestion, Don't Build More Roads—Close 'Em." *Corvallis Gazette-Times* (Corvallis, OR) May 20, 1998.
35. Ibid.
36. Ibid.
37. James Howard Kunstler. "The War on Cars." *Slate* January 21, 1998. http://www.slate.com/articles/news_and_politics/dialogues/features/1998/the_war_on_cars/_3.html (accessed February 21, 2013).
38. Bishop, "To Reduce Congestion."
39. Chris Maser, Russ Beaton, and Kevin Smith. *Setting the Stage for Sustainability: A Citizen's Handbook.* Lewis, Boca Raton, FL, 1998, 275 pp.
40. Patricia Mulder. "Time to Resume Our Dialogue on Growth." *Corvallis Gazette-Times* (Corvallis, OR) May 10, 1999.
41. Ibid.
42. Aaron Corvin. "Corvallis Population Hits 50,000." *Corvallis Gazette-Times* (Corvallis, OR) July 7, 1999.
43. The discussion of Seaside, Oregon, is based on the following article: Associated Press. "Seaside Leaders Focus on Environment." *Corvallis Gazette-Times* (Corvallis, OR) July 12, 1999.
44. Ibid.
45. Ibid.
46. The following discussion of carrying capacity is based on Maser, *Sustainable Community Development.*
47. Garrett Hardin. "Cultural Carrying Capacity." 1986. http://www.garretthardinsociety.org/articles/art_cultural_carrying_capacity.html (accessed February 21, 2013).
48. Ibid.
49. David Skrbina. "Convivial Communities." *Resurgences* 196 (1999):16–18.
50. Ibid.

6

Reframing the Problem

> We are losing our living systems, social systems, cultural systems, governing systems, stability, and our constitutional health, and we're surrendering it all at the same time.
>
> **—Paul Hawken[1]**

> If we directly face the pain of what's here in our midst, we won't become blinded by blame. Past the pain is the possibility of clear-seeing, unclouded by despair or paranoia. From here, we can see the way forward together.
>
> **—Margaret Wheatley[2]**

Human strife, species extinction, vanishing resources, climate change, and other persistent and alarming trends move many to action or a desire to act, even when basic necessities are being met and freedom, good health, and a sense of safety are intact. Fear frequently occludes a sense of well-being, which propels many to act in anticipation of systems collapse, believing that the inevitable can be delayed, mitigated, or perhaps even diverted. While some people despair and see only the darkness, others are excited about the possibilities emerging from the need for triage and the creativity that can be forged by the heat of urgency.

In such a context, confusion and uncertainty can dominate, yet, to be effective, our need for certainty must be managed. As Margaret Wheatley so aptly puts it, "Certainty is a very effective way of defending ourselves from the irresolvable nature of life" (p. 15). But, simply staying present to "what is" helps us see why optimism and pessimism keep us from "fully engaging with the complexity of this time." "If we see only troubles, or only opportunities, we are, in both cases, blinded by our need for certainty, our need to know what's going on, to figure things out so we can be useful" (p. 15). Wheatley goes on to suggest that to "dwell in the uncertainty, hold the paradoxes, live in the complexities and contradictions without needing them to resolve" is a much more powerful place from which to create change.[3] We (Jane and Chris) believe it is from this place of nonjudgment that the path toward sanity, clarity, and ingenuity can be known, and while the fear may be unavoidable, we will be in a better position to choose how and when to respond. It is from this place of nonjudgment that we can reframe the problem and act with integrity, imagination, and creativity.

Much of the literature addressing the collapse of human-made and natural systems, societal collapse, social injustice, "phantom wealth,"[4] and so on dedicates huge amounts of space to such human character defects as greed, lust for power, anger, xenophobia, and racism, among others. Rarely, if ever, is there offered a framework for viewing destructive human behavior and mindsets that, if not disarm, mollify, our disgust, rage, anger, or fear—reactions that cause in us the very behaviors we protest. Therefore, we offer a few ways to disarm such negativity insofar as we subscribe to the belief that it is "hard to be angry and creative at the same time."[5] Moreover, harnessing positive, human creativity in solving planetary problems is absolutely critical at this time because, as Albert Einstein said, "We can't solve problems by using the same thinking we used when we created them."[6]

We humans introduce thoughts, practices, substances, and technologies into the environment, and we usually think of those introductions in terms of development. Whatever we introduce into the environment in the name of development will consequently determine how the environment will respond to our presence and to our cultural necessities. It is therefore to our social benefit to pay close attention to what we introduce because, once introduced, it is forever out of our control.

The safe disposal of nuclear waste is a prime example. Consider that, as of 2012, there were 6 underground tanks—of 177 aging tanks storing millions of gallons (liters) of radioactive sludge at the Hanford Nuclear Reservation in Hanford, Washington—leaking. Moreover, it is estimated that the six tanks are leaking around 1,000 gallons (3,780 liters) of the radioactive sludge annually,[7] the fluid of which will leach into the groundwater and eventually into the nearby Columbia River, which will carry it into the Pacific Ocean—contaminating everything along the way.

Introduction of a foreign substance, process, or technology into the environment often has a much greater impact on an ecosystem's ability to function the way we want it to over time than we humans have anticipated. The things we introduce into the environment represent our values. After all, values—often competing values to which we give no conscious thought—shape the contours of our lives, which raises the question of how one goes about calculating the risk of something that has never before happened and, by our reckoning, is unlikely to occur in the future. Nevertheless, as technology compounds technology—all designed to facilitate our human sense of control, the question becomes, can we control the effects of that which we introduce into the environment?

As society develops new technology, draws on the resources of Earth, and generates unprecedented quantities of unintended industrial products, such as toxic wastes, this question is being asked more and more frequently because we must understand, as best we can, what the effects of our activities will be on the sustainability of our environment as a whole. Some people would throw their hands up and say that such predictions are impossible, but that is not entirely true. Over the past few decades, says retired physicist and engineer B. John Garrick, an entire discipline, known as risk assessment or risk analysis, has been formulated around the proposition of "what if."

As such, assessment of risk focuses on three issues, according to Garrick: "What can go wrong? How likely is it? And, what are the consequences?"[8] The answer is given as a probability, which is the language of uncertainties, and learning how to quantify the uncertainties is a critical part of assessing risk.

Many people do not find the probabilities reassuring, however, despite how low they may be. I (Chris) am one of those people because no matter how low they may be, I know how ignorant we are collectively when it comes to the synergistic effects of interdependent living systems. My doubts are not based on fear, however, but rather concern for the expressed surety of our human knowledge. Nevertheless, every decision has a risk attached to it, and the more we can understand the risks of our proposed actions, the better off we will be.

Making people more comfortable with the probabilities, says Garrick, is a matter of changing the terms of the debate. He goes on to say: "During my 40-plus years in the risk [assessment] business, the questions that have come to annoy me most are 'How safe is safe?' or 'How much risk is acceptable?' These are illogical questions. The only answer that makes sense is 'It depends—on the alternatives available and on the benefits to be gained by making a certain decision.'"[9]

The questions are not illogical to me, as they are to Garrick, because the questions, as asked, simply point out that the people who ask them are frightened and do not know how to frame the questions in a way that addresses their fears. In addition, they are questions of value, including things likely to be lost, such as some long-cherished, often-intangible component of one's lifestyle, which is far more complicated to deal with than the simple, traditional, linear questions asked and favored by scientists and engineers, who can at best measure only tangible effects.

Be that as it may, the best possible assessment of potential risk, as Garrick states, requires participation by the public, either directly or through elected representatives. "Governments and the private sector," admonishes Garrick, "need to develop mechanisms to ensure this input [by the public] without letting the process get bogged down by a few people whose entire agendas may never be expressed and whose actions lead to gross mismanagement of society's resources. Those who spread false information [on all sides] need to be held accountable—especially since the consequences of their actions can cost billions of dollars,"[10] to say nothing of degraded ecosystems and the loss of the services they provide, which add to the misery of countless people.

"If decisions involving risk are not approached rationally, they will be made on political and emotional bases, which usually is not optimal for society," counsels Garrick.[11] But then, all truly *rational* decisions involve—but are not ruled by—emotions, which after all are the foundation of our values as human beings. Politics, on the other hand, is often a case of hidden agendas that determine who wins and who loses. Assessing risk in a formal manner provides us with a way to better understand the possible consequences of the choices we face.

We would be well advised, therefore, to make the best possible use of what we know about assessing risks, provided we do not become enamored with the outcome as a sure bet. And, a bet on the outcome of adding something to or subtracting something from an ecosystem is all risk assessment is.

When we conduct a risk assessment, we are acting mindfully, which does not come easily for many. Although we know that research on human behavior includes findings regarding our capacity for love, compassion, and altruism, we question, as do many, why humans seem so insistent on destroying life and the systems that support life. Perhaps more will be known as a result of President Obama's recently announced initiative to "advance the knowledge of the brain's billions of neurons and gain greater insights into perception, actions and, ultimately, consciousness."[13] Granted, the project is motivated by known, debilitating diseases, such as Alzheimer's and Parkinson's, but it may add to our knowledge regarding why humans contribute so significantly to their own demise. What, we wonder, are some of the possible subconscious drivers of planetary destruction?

All of Life Is Cyclic

We can say with relative certainty that, like everything else, we are part of a cycle from birth to death and beyond, and everything we create, including culture and institutions, is likewise cyclical. Everything is in a constant process of change, of becoming something else, which consists of eternal, irreversible novelty, such as birth, growth, maturation, decline, demise, and beyond.

Consider that civilizations have evolved by similar steps: growth of intelligence through discoveries and inventions, through the ideas of government, family, and property, all based on a slow accumulation of experimental knowledge. As such, civilizations have much in common, and their evolutionary stages are connected to one another in a natural sequence of cultural development.

The arts of subsistence and the achievements of technology can be used to distinguish the periods of human progress. People lived by gathering fruits and nuts; learned to hunt, fish, and use fire; invented the spear and atlatl; and then the bow and arrow. They developed the art of making pottery, learned to domesticate animals and cultivate plants, began using adobe and stone in building houses, and learned to smelt iron and use it in tools. Finally, what we call "civilization" began with the invention of agriculture and concomitant settlement, the phonetic alphabet, culminating in all the current-day knowledge and technological wonders.

Each civilization has also been marked by its birth, maturation, and demise, the last brought about by uncontrolled population growth that outstripped the source of available energy, be it loss of topsoil, deforestation, or the forced degradation of nature's biophysical processes and thus the demise of nature's biophysical services. But, in past eras, the survivors could move on to less-populated, more fertile areas as their civilizations collapsed. Today, there is nowhere left on Earth to go.

Yet, having learned little or nothing from history, our society—and humanity at large—is currently destroying the very environment from which it sprang and on which it relies for continuance. Human society, as we know it, cannot be the final evolutionary stage for human existence. But, what lies beyond our current notion of civilization? What is the next frontier for "civilized" people to conquer? Is it outer space as so often stated? No, it is not outer space. What then? It is inner space, the conquest of oneself, which many assert is life's most difficult task.

In the material world, self-conquest means bringing one's thoughts and behaviors in line with the immutable biophysical principles governing the world in which we live (see Chapter 3). In the spiritual realm, this means disciplining one's thoughts and behaviors in accord with the highest spiritual/social principles handed down throughout the ages, such as love your neighbor as yourself and treat others as you want them to treat you.

The outcome of self-conquest is social-environmental sustainability, which must be the next cultural stage toward which we struggle. Social-environmental sustainability is the frontier beyond self-centeredness, and its stepchild, destructive competition and conflict, which destroys human dignity, degrades an ecosystem's productive capacity and thus forecloses options for all generations.

To fulfill our acknowledged obligation to future generations requires fundamental changes in our social consciousness and cultural norms, changes that will demand choices different from those we have heretofore made, which means thinking anew.[14] But, "a great many people," as American psychologist William James observes, "think they are thinking when they are merely rearranging their prejudices."[15] That said, if we look at historic patterns of civilizations as have authors Joseph Tainter and Gerod Diamond,[16] we will have a sense of some persistent human traits that lead to the demise of cultures and society as a whole.

In terms of a human strategy for acquiring and maintaining power, Machiavelli's *The Prince*,[17] written in the sixteenth century, leaves little doubt that a combative strategy is alive and well in the twenty-first century and could possibly be seen as second nature for humans.

With some things, we can easily observe nature's cycles as the sun traverses the sky, in the flow and ebb of the seasons as flowers come and go in our gardens, in the aging of our bodies, and even in our communities and civilization itself, as attested by the foregoing discussion. We know, for example, that death is imminent with our bodies, and we are either certain or uncertain about what happens next, if we are thinking about it at all. Therefore, in the most neutral sense, all life is on the same journey to the same end, and to hold this frame of reference may serve, at the very least, to mitigate our adverse reaction to what we perceive as the human tendency toward self-destruction. To release judgment and live in the present moment—in the here and now, which is all we ever have—is a choice, our choice to give up fear of death and the need to control the uncontrollable. And, if we choose wisely, we can shed judgment and thus address our problems more effectively.

We Make What We Are

Profound and significant change to social and economic structures wrought by humans is most obvious in the realm of technology. More often than not, technological advances dominate the news. Advances in information technology that seem exponential in nature impel society forward. We are astoundingly prolific in this realm. It seems we make what we are.

In Steven Levy's review of two books published in 1993—The *Fourth Discontinuity* and *Metaman*, Levy ponders the following: "Could human creations eventually claim equality with us? Would we regard them as partners? More than 100 years ago, Samuel Butler had a character in his novel 'Erewhon' utter a startling prophecy: 'There is no security against the ultimate development of mechanical consciousness.' Yet, as the authors of two new books proclaim, we need not wait for indisputably intelligent machines to close out the age of biological autonomy." According to *The Fourth Discontinuity* and *Metaman*, the barrier between humans and machines has already evaporated.[18]

Might we posit as well that we are so insistent on technological innovation because the technology we create is simply a natural extension of our own "wiring"? In being wired as we are—designing and building technology to augment the abilities of our brains and simultaneously "make what we are"—we create greater complexity, just as we can see in nature when an animal evolves from a single cell to a highly differentiated cellular organism. Increasing complexity therefore is part of the cyclical nature of life: from the simple to the complex and back to the simple with demise and decay. This can also be observed in the institutions and multiplicity of roles we create.

Humans have surrounded themselves with artifacts that have radically transformed our environment. In so doing, we have created systems that alter the structures of nature on which we rely for survival. Thus, we unwittingly add to our growing uncertainty about the conditions of our lives.[19] As we create change, we add to complexity, even though we have serious limitations when it comes to managing complexity, including trade-offs. Therefore, even if we take actions for the greatest good for the greatest number, we will fail due to our lack of considering the irreversible effects of what we introduce into the environment, to say nothing of our inability to know long-term outcomes.[20]

The creation of complexity therefore comes with societal and environmental costs. According to Joseph Tainter, our problem-solving techniques frequently involve the creation of new layers of bureaucracy, infrastructure, or even a social class to address a current challenge, which in turn and requires an "energy" subsidy or the utilization of resources or other forms of wealth. Tainter argues that sustainability or collapse of societies follows from the success or failure of problem-solving institutions or when investments in social complexity and the necessary "energy subsidies" reach a point of diminishing marginal returns. He frames "collapse" as a point when a society is forced to rapidly shed a significant portion of its complexity.[21]

Thus, to return to the notion that we are what we make, we can see our existence, which is cyclical, exported to what we make and that both involve expanding complexity and ultimate collapse. To understand where we are in the cycle is critical to adaptability, agility, and the quality of innovation. Tainter goes on to advise as follows:

> Regardless of when our efforts to understand and resolve contemporary problems reach diminishing returns, one point should be clear. It is essential to know where we are in history. If macroeconomic patterns develop over periods of generations or centuries, it is not possible to comprehend our current conditions unless we understand where we are in this process. We have the opportunity to become the first people in history to understand how a society's problem-solving abilities change. To know that this is possible yet not to act upon it would be a great failure of the practical application of ecological economics.[22]

If we are to advance sustainability, therefore, we need to understand our limitations in managing complexity, know where we are in history and cycles, and continue to enhance our ability to think systemically. In the realm of land use planning, we need to plan at a comprehensible scale and, as suggested by Van der Rym and Cowan in *Ecological Design*, learn how to link scales. "Nature's processes are inherently *scale linking*, for they intimately depend on the flow of energy and materials across scales." Therefore, scale linking "reminds us of the wider environmental consequences of our designs. Unless we work with nature's own finely tuned scale-linking systems, we endanger the stability of life on this planet" (p. 51).[23]

Humankind in Amnesia

One of the more interesting hypotheses regarding why humans seem insistent on self-destruction comes from the writings of Immanuel Velikovsky

(1895–1979), a Russian psychiatrist and independent scholar.[24] Velikovsky, vilified by peers for his radical ideas, is best known as the author of a number of controversial books, in particular the U.S. bestseller *Worlds in Collision*, published in 1950, reinterpreting the events of ancient history.[25]

Using comparative mythology and ancient literary sources, which include the Old Testament, Velikovsky argues that, in ancient time, Earth suffered catastrophic close contacts with other planets, mainly Venus and Mars. By coupling these close contacts with the notion of racial memory,[26] which translates into "feelings, patterns of thought, and fragments of experience that have been transmitted from generation to generation in all humans and have deeply influence the mind and behavior," Velikovsky poses a possible way of perceiving the human tendency toward ignoring or contributing to systems collapse, a trajectory that few would argue is not apparent.

In his 1982 book *Mankind in Amnesia*, Velikovsky draws from the work of Freud and states that the racial memory of cataclysmic events on the planet may "pilot the personality toward its bizarre behavior and, not seldom, toward a repetition of the traumatic experience" (p. 20).[27] Might the concept of racial memory be a nonjudgmental way of viewing the brutality and violence of war as behavior driven by the subconscious memory of some cataclysmic event on Earth? If so, could this be a framework for disarming our fears, and their counterparts of blame, to find innovative, systemic solutions for today's global social-environmental crises?

Competing Instincts and Ecological Unconscious

Most of us are painfully aware of how seriously humans have compromised the biophysical integrity of the planet by competing for, and thus depleting, its resources for short-term gains. Some people believe such pain—usually expressed as a fear of losing something considered to be of personal value—is part of an ecological unconscious and therefore part of some basic human wiring.

The growing field of "ecopsychology" embraces the following paradigm: "Just as Freud believed that neuroses were the consequences of dismissing our deep-rooted sexual and aggressive instincts, ecopsychologists believe that grief, despair and anxiety are the consequences of dismissing equally deep-rooted ecological instincts."[28] Further, ecopsychology holds that "there is a synergistic interplay between planetary and personal well-being … [and that] the needs of the planet are the needs of the person, the rights of the person are the rights of the planet."[29] Evidence of this deeply rooted connection or instinct frequently shows up when people experience a sense of loss associated with such things as clear-cutting a forested hillside that contains one's much-loved hiking trail; damming a river, thereby destroying its long-cherished scenic beauty; subjecting one's favorite neighborhood flower-bedecked meadow to an absentee-owner housing development—all actions that scar or make a significant change to Earth and one's sense of place.

Glen Albrecht coined a term to capture this state of being: "solastalgia," a combination of the Latin word *solacium* ("comfort") and the Greek *algia* ("pain"), which he defines as "the pain experienced when there is recognition that the place where one resides and that one loves is under immediate assault ... a form of homesickness one gets when one is still at 'home.'"[30]

We know, in other contexts, that people have competing instincts. Take, for example, endangering one's own life in the process of protecting a child from harm. The question, therefore, is what instinct prevails or drives us to put our own life support system in jeopardy if, indeed, "deeply rooted ecological instincts" exist? Is there other, competing, subconscious circuitry at work about which we need to be curious? We think it would be prudent to do so.

According to Robert Port, professor of linguistics and cognitive science, Indiana University, "One of the challenges of each human's life is the reconciliation of all ... competing instincts. This is where individuals exhibit the greatest differences and where communities, cultures and families may bias our behavior in distinctive ways."[31]

How we "manage" our instincts requires first acknowledging they exist and then observing how they manage us. Whether we take further action is then simply a choice.

Removing the Blame Factor

Finally, we see sustainability as a blame-free model. Simply put, if everything is connected to everything else in an interdependent way, and change is a constant, irreversible process of eternal novelty, how can we logically point the finger of blame at any one individual or entity? And, more important, what good does blaming one another do? Whereas we may think we know the sole perpetrator, as in a dictator driving a genocidal war, there remains the question of what underlies his or her psychosis of fear. What is the historical context, both personal and societal?

Clearly, we will never fully understand the infinite novelty uniting the systems with the systems that comprise our home planet. Yet, if we take the time to reflect on these infinite connections, we can harness the creativity needed for developing systemic solutions to the complex problems of social-environmental sustainability that are today plaguing our global human society and thus benefit all generations.

Summary

Chapter 6 unravels barriers to creativity in the planning process by discussing some possible reasons for them that seem embedded in human nature and how to overcome them by reframing the problem in a more positive way. Chapter 7 introduces the importance of and ways of modeling the planning process after nature. Modeling land use planning after nature is a vital step in our human journey toward social-environmental sustainability.

Endnotes

1. Paul Hawken. http://www.brainyquote.com/quotes/authors/p/paul_hawken.html#8zfuJamaAER552dh.99 (accessed January 18, 2013).
2. Margaret J. Wheatley. *Perseverance*. Berrett-Koehler, San Francisco, 2010.
3. The foregoing discussion is based on Wheatley, *Perseverance*.
4. David Korten. *Agenda for a New Economy: From Phantom Wealth to Real Wealth*. Berrett-Koehler Press, San Francisco, 2009.
5. Paul Hawken. Northland College commencement speech, Ashland, WI, June 1996.
6. Albert Einstein. http://www.thequotefactory.com/all/quotes?author=einstein&max=10&maxsteps=10&offset=80) (accessed January 17, 2013).
7. (1) Shannon Dininny. "Gov: 6 Underground Hanford Nuclear Tanks Leaking." http://abcnews.go.com/US/wireStory/gov-underground-hanford-nuclear-tanks-leaking-18570507 (accessed February 28, 2013); and (2) Mike Baker. "Inslee Says Hanford Leaks Could Be 1,000 Gal/Year." http://abcnews.go.com/US/wireStory/inslee-hanford-leaks-1000-galyear-18611537 (accessed February 28, 2013).
8. The following comments on risk are gleaned from B. John Garrick. "Society Must Come to Terms with Risk." *Corvallis Gazette-Times* (Corvallis, OR) November 9, 1997.
9. Ibid.
10. Ibid.
11. Ibid.
12. Einstein, Endnote 6.
13. John Markoff. "Obama Seeking to Boost Study of Human Brain." *The New York Times* February 17, 2013. http://www.nytimes.com/2013/02/18/science/project-seeks-to-build-map-of-human-brain.html?pagewanted=all&_r=0 (accessed February 21, 2013).
14. The foregoing discussion of civilization is based on Chris Maser. *Resolving Environmental Conflict: Towards Sustainable Community Development*. St. Lucie Press, Delray Beach, FL, 1996.

15. William James. http://www.searchquotes.com/quotation/Many_people_think_they_are_thinking_when_they_are_merely_rearranging_their_prejudices./20447/ (accessed February 28, 2013).
16. (1) Joseph Tainter. *The Collapse of Complex Societies*. Cambridge University Press, New York, 1990; and (2) Jared Diamond. *Collapse: How Societies Choose to Fail or Succeed* (revised edition). Penguin Books, New York, 2011.
17. Nicolo Machiavelli. *The Prince*. http://www.gutenberg.org/files/1232/1232-h/1232-h.htm (accessed February 28, 2013).
18. Steven Levy. "We Are What We Make." *The New York Times* October 24, 1993. http://www.nytimes.com/1993/10/24/books/we-are-what-we-make.html (accessed December 30, 2012)
19. Braden Allenby. Presentation to the Institute for Human, Machines and Cognition. "Life on a Terraformed Planet: Please Fasten Your Seatbelts for Takeoff." October 25, 2007.
20. Ibid.
21. Joseph A. Tainter. *Getting Down to Earth: Practical Applications of Ecological Economics*. Island Press, Washington, DC, 1996.
22. Joseph A. Tainter. "Complexity, Problem Solving, and Sustainable Societies." http://sep.rockefeller.edu/wp-content/uploads/2011/02/Tainter_Complexity-Problem-Solving-Sustainable-Societies_1996.pdf (accessed June 4, 2013)
23. Sim Van der Rym and Stuart Cowan. *Ecological Design* (10th anniversary edition). Island Press, Washington, DC, 2007.
24. "Immanuel Velikovsky." http://en.wikipedia.org/wiki/Immanuel_Velikovsky (accessed December 4, 2012).
25. Immanuel Velikovsky. *Worlds in Collision*. Doubleday, Garden City, NY, 1950.
26. "Racial Memory." http://dictionary.reference.com/browse/racial+memory (accessed December 5, 2012).
27. Immanuel Velikovsky. *Mankind in Amnesia*. Doubleday, Garden City, NY, 1982.
28. Daniel B. Smith. "Is There an Ecological Unconscious?" *The New York Times* January 27, 2010. http://www.nytimes.com/2010/01/31/magazine/31ecopsych-t.html?pagewanted=all&_r=0 (accessed December 30, 2012).
29. Theodore Roszak. "Awakening the Ecological Unconscious." 1993. http://www.context.org/iclib/ic34/roszak/ (accessed January 5, 2013).
30. Ibid.
31. Robert R. Port. "Possible Human Instincts." April 28, 2000. http://www.cs.indiana.edu/~port/teach/205/instincts.html (accessed January 5, 2013).

7

Modeling the Planning Process after Nature

> Democracy is a non-stop experiment in the strengths and weaknesses of our political institutions, local communities, and the human heart—and its outcome can never be taken for granted. The experiment is endless, unless we blow up the lab, and the explosives to do the job are found within us. But so also is the heart's alchemy that can turn suffering into compassion, conflict into community, and tension into energy for creativity amid democracy's demands.
>
> —**Parker J. Palmer**[1]

In community planning, the process, which creates, implements, and periodically revisits plans, is ideally modeled after nature's biophysical patterns, systems, and cycles. People conduct such a process to ensure their future, as much as humanly possible, in an ecosystem of which they are an inseparable part. Therefore, when modeling a planning process, one must consider the multidimensional biophysical requirements of both the people and the ecosystem. The multidimensionality of people means that conflict can be anticipated. The challenge is to use conflict, as necessary, to fuel change, and as Palmer suggests, fuel the community.

The framework of sustainability for community development rests, in part, on the definition of *sustain*, which is "to support or hold up from below" and also "to keep a person's mind and spirit from giving way."[2] This framework recognizes the importance of meeting the complex necessities for human survival while maintaining the integrity of biophysical systems that support all life. Sustainable development is intended to make it possible for future generations to have as many, if not more, options than the current generation—options for the nurturance of body, mind, and spirit.

A more sensitive, multidimensional approach to sustainable development recognizes that the well-being of body, mind, and spirit is inextricably linked to the human context in and of the natural world. If a social/legal system, such as municipal government, is concerned only with the minds and intellects of its constituents, while failing to recognize the full range of human necessities and their embeddedness in the rest of nature, the usefulness and functionality of that social/legal system will ultimately devolve. Today's wars and the resultant human chaos and the collapse of governments and social systems is evidence of the failure on the part of leaders to understand how essential it is to both recognize *and* accommodate the health and inner growth of the whole human being. Moreover, deteriorating ecosystems also signal failure of governments to recognize and adequately address the exceedingly complex, multidimensional nature of the biophysical system on which humans, as inseparable components, rely for physical, emotional, and spiritual well-being.

In sustainable communities, accountability for the integrity of the environment, the health of the economy, and the equity of human relationships characterizes the decision-making process. Such a multidimensional process is modeled after the biophysical principles that govern nature.

Within such a framework, land use decisions respect several ecological dimensions associated with time, interrelationships of species, natural disturbances, and the connectivity of habitats.[3] Chapter 5 elaborates on this type of land use management by emphasizing the importance of moving purposefully, consciously toward the connectivity of habitats in the form of well-planned and protected systems of open space. Habitat, kept intact, provides the shelter and food for other creatures; provides ecosystem services, including water catchment, purification, and groundwater recharge; and assists in the maintenance of biological and genetic diversity. Planning for change in a sustainable community includes the protection of habitats through the protection of open spaces. For sustainable community to exist, however, this action must be linked with other actions that also lead toward sustainability.

In modeling our planning process after nature, we can look to the same characteristics of the initiatives, policies, and processes identified as part of a local living economy in Chapters 4 and 5. These examples represent efforts that acknowledge and seek to align with biophysical principles by (1) achieving zero waste; (2) creating and maintaining diversity within the planning process; and (3) strengthening the flow of energy through self-organization.

Zero Waste

A zero waste paradigm implies reconceptualizing waste. This means recasting waste as something potentially useful, much like recasting conflict as necessary fuel for change. John Jackson, former chairperson of Great Lakes United, a U.S.–Canadian environmental coalition, refrains from using the term *waste* and uses instead the phrase "valuable used materials." His report to the government of Ontario, "Resources, Not Garbage," admonishes the provincial government to reconceptualize waste to protect valuable resources, decrease the use of energy, and decrease toxic contaminants at the stage of production, as well from waste disposal facilities, thereby protecting not only air quality but also surface and groundwaters.[4]

In a municipal planning process, zero waste can be achieved by openness to diverse opinions and ideas, efficient *and* effective use of time through carefully planned and facilitated meetings, and effective use of technology (such as remote access) to maximize participation, as well as meeting attendance. This approach will allay a common citizen complaint that planning is a waste of taxpayer dollars, evidenced by plans sitting idly on the shelves of planning departments, where their greatest utility is to gather dust. A complaint of this nature is a call for education—not defense. Furthermore, it must be viewed as an opportunity—instead of an obstacle—and thus another way of reconceptualizing a negative into a positive.

If all experiences are seen as useful, they can no longer be viewed as a waste of time. Even if a city undertakes an expensive planning process, which leaves the community without a vision or a plan, some learning will have taken place—provided negativity does not override the collective ability to move forward in spite of what might be viewed as a setback. Learning experiences can sometimes be exceedingly costly, in terms of both time and money, but if used as educational stepping-stones, neither time nor money is truly lost.

Diversity within the Planning Process

A basic, biophysical principle is that diversity is directly related to ecosystem resilience, which, from a human point of view, equates to system stability. To the extent that diversity is diminished, so the stability, integrity, strength, resiliency, or durability of any system is diminished.[5] Business and industry are learning this principle, as evidenced by the ever-increasing number of corporations diversifying their interests and activities. Built-in backup components or "subsystems" are part of diversity and thus a system's resilience—if one part fails, another is available to serve the same function. For example, creating a place in the planning process for all voices and the safety necessary for free expression is unlikely to capture all the possible ideas. However, what is frequently said in parking lots after public meetings or in hallways or restrooms—that which Peter Senge calls "left column thoughts"[6]—can be lost if not spoken to the larger group. Because such thoughts may be critical to a fully informed, realistic, and meaningful vision and plan for a community, it is wise to request that people write down the ideas resulting from such discussions and make them available to the whole group.

In light of this, it is wise to engage people with divergent ideas and political views in the planning process, people of varying ages and incomes, race and ethnicity, cultures, education, and religious and sexual preference. In this way, we conserve, rather than waste, human experiences. With such a diverse group, conflict is more likely to surface, which some may view as negative and thus a waste of time. Disagreement with civility is at times necessary, however, to arrive at a consensus and must be viewed as both a catalyst and fuel for change.

Inclusivity is a means of maintaining diversity of participation, thought, opinion, and expression. Although a facilitator of a planning process can take the lead in modeling and maintaining inclusive behaviors, the group's acknowledgment of certain principles at the outset is beneficial. Further, it is a good idea to post them for all to see as a reminder of the agreed-on rules of civility in the planning process. The following list represents those principles we have found useful in the work we have done with communities:

1. Remain open to a continuum of thought and opinion and the notion that everyone is right from his or her perspective; this means giving up seeing another's view as right or wrong, rather as somewhere on the continuum of rightness. In nature, stability and diversity are highly interrelated.
2. Understand that the quality of the questions are more important than the answers because all an answer ever does is redefine the question, to which Alfred North Whitehead offers this counsel, "The art of progress is to preserve order amid change and to preserve change amid order."[7]
3. Stay present to what is. Full acceptance—without judgment—will free both personal energy and creativity. Margaret Wheatley suggests: "Be present to what is, accept what is, and there you will find the corridors of power."[8] And, when possible, choose to believe that there are "no problems, only solutions."[9]
4. View everyone as a designer with important ideas, perspectives, and information to contribute.
5. Accept our ignorance while trusting our intuition and doubting the certainty of our knowledge.[10]
6. Acknowledge that the well-being of another is a measure of our own well-being.
7. Embrace traditional knowledge, which coevolves from culture and place and is a critical asset.[11]
8. Remember that common welfare must come first, and that power resides in unity of the group only.
9. Seek a solution space for design unbounded by simplification or ideology; evaluation of design solutions considers scale, scaling, and emergent events across foreseeable timelines.[12]
10. Above all, hold principles above personalities.

Strengthening the Flow of Energy through Self-Organization

Resiliency and diversity characterize a process modeled after nature, as is self-organization—as within a community's set of values—to adapt to ongoing environmental change. Self-organization, which strengthens the flow of energy, is inherent in all living systems. In biological systems, it is called evolution. In human economic systems, it takes the form of technological advance or social revolution. Self-organization means changing any given aspect of a system, such as adding or deleting new physical structures, negative or positive feedback loops, flows of information, or rules. The ability to self-organize is the strongest form of resilience in a system; it is the ability to survive change through the flexibility of the response.[13]

Self-organization occurs many times following natural disasters. In 1989, Santa Cruz, California, experienced the effects of an earthquake that measured 7.1 on the Richter Scale. I (Jane) was a planner for the city at that time. What followed this event was similar to the community's response to several other natural disasters occurring in Santa Cruz over a 12-year period. The coming together of a community when some event affects all citizens, especially when the impact creates the need for mutual assistance, demonstrates, in our opinion, the phenomenon known as self-organization. A primary characteristic of the resulting social system is its diversity.

The similarity of experience, the shock and horror resulting from the earthquake's lift and surge of enormous force, brought the community together. It is during events like these when age, race, income, social status, and religion lose their divisiveness. The humanity of each person coming forward to help was all that mattered. Everyone seemed to be giving what he or she could. And, the level of acceptance for anything anyone had to offer was high.

The story of the earthquake in Santa Cruz brings to mind a tune, "Hell and High Water." Written by Warren Nelson for a historical musical about the town of Bayfield, Wisconsin, a small fishing village on the south shore of Lake Superior, the musical tells of a huge rainstorm and resulting flood. This poignant song illustrates that, "It seems it takes disaster and emergency, to show a town it's just one big family."[14]

Re-creating the type of self-organization that occurs in a community after a natural disaster, without the catalyst of a disaster itself, is one goal of the planning process we are doing our best to describe in this book. How can this be accomplished? A place to start is to characterize the catalyst in ways that might translate into the planning process.

To begin, let us note that events besides natural disasters have drawn communities together in ways that do not occur otherwise. An example is the recent tragedy in Newtown, Connecticut, where 26 people, mostly children, died as a result of a young man opening fire in the school building. What characterizes these events that create a sense of belonging, equality, mutual suffering, usefulness, cooperation, and place?

First, the event touches everyone in the community in some way, such as a sense of mutual suffering. Second, the event is such that someone can easily empathize with the heart-breaking experience of another person. Third, the event affects some individuals more dramatically than others; therefore, some need more assistance than others. Fourth, there is recognition that everyone has something to offer in the recovery process, if only a smile. Fifth, it is clear that the event could reoccur to virtually anyone, anytime, anywhere. In the case of an earthquake, both aftershocks and the possibility that the major tremor could be a precursor to something greater makes a huge contribution to the sense of "community." Sixth, there is a sense of urgency. The time to act is now, which blurs the sense of tomorrow in the immediacy of the eternal moment—the here and now, which is truly all we ever have.

How can the planning process in a community re-create this type of community cohesion? First, the certainty of dramatic, impending change and its eternal novelty is the "event" that initiates the planning process, whereas the uncertainty of impending change that hangs over a community if it chooses not to plan is the reality that will affect everyone and will affect some in ways that are more dramatic than others.

The certainty of change, the uncertainty of its outcome, and the fact that change is constantly occurring in novel—irreversible—ways may be one way of engaging the community. The invitation to become involved can be done in a way that emphasizes the importance of the contribution of all. How? This occurs by inviting all to share their vision—based on their dreams—for the future of their place. Most people in a community care about how the future of their place of residence or business unfolds.

But, what, one might ask, is the urgency? Why do we need to plan now? And, most important, how does the agency or body initiating a planning process do so in a way that draws people into the spirit of community without relying on fear as the catalyst or the "hook"?

One might, for example, point out to a community in an open invitation through news stories or in open, preplanning meetings that changes are occurring, that the impending changes will affect some citizens more than others, and that, without careful planning, the results of such changes could be a heightened sense of uncertainty, especially if there is no strong organizing context of a shared vision to guide the collective future of the community.[15] Because people generally care about their place, the invitation is based on love for the community—not on the fear of change.

A celebration could be planned to launch the planning process, one that highlights the reasons people choose to live where they do. The fragility of everything that a community holds dear could help to bring poignancy to this kickoff event—a town's history, its natural wealth and beautiful features, its downtown, its favorite gathering places—all enshrined in premeeting lists compiled by attendees who have walked around their community recording what they love about it and want to protect.

At one postearthquake charette in Santa Cruz (*charette* is a condensed community design process led by design professionals), for example, the first activity of attendees as they entered the high school gymnasium was to show where they lived by sticking a pin in the city map. Second, after breaking into smaller groups to discuss the possibilities for the recovery of Santa Cruz's downtown, where about 60 buildings were seriously damaged or destroyed, each participant was asked to draw a heart where they felt the "heart of downtown" was. The same kind of thing can be done to engage people without having first suffered a disaster.

It is important to emphasize in the invitation that everyone's perspectives, ideas, feelings, and willingness to share their time and energy is both needed and appreciated. Questions not only need to be asked but also are an excellent place to begin the planning process: Why is our community experiencing these changes? Why did we not notice them before they reached this stage? How has growth changed our community? What are the options for addressing such issues as growth? In light of such questions, it must be repeatedly stressed that we, as citizens, need to help one another to protect that which we value in this, our tiny place on Earth.

Free-Flowing Communication

For communication to take place, four elements need be present: (1) the source of the message, (2) the message itself, (3) the receiver of the message, and (4) the medium through which the message is disseminated. In such a model, there is the possibility of "noise" between the source and the recipient, which alters the original message. In healthy systems, communication and the flow of information (in whatever form it is perceived) are constant and unobstructed. Interruption or distortion of the message can block necessary, built-in feedback loops, which reduces or stops the flow of information. When communication is interrupted, all components of a system are affected. The healthy pulse of a human heart exemplifies a steady source of information to the organism. Blocking one of the arteries in this process can have widespread and sometimes lethal results.

Human communication is comprised of motives, attitudes, thoughts, words, and actions. Thus, a planning *process* communicates to the public how the entity initiating the process (for example, the mayor or the state legislature) feels toward the public. It reflects an attitude that communicates more directly than words and more loudly, as we are reminded by Ralph Waldo Emerson: "Your attitude thunders so loudly I can't hear what you say."[16]

Historically, many city plans are brought to the public for approval *after* their development, which communicates a number of things to the citizens, including (1) the planning staff is already attached to and committed to a plan; (2) negotiation of the plan has already taken place; (3) the entity that initiated the plan only wants the citizens to approve it as written; (4) the experts recognized by the entity that initiated the plan are the only ones with valid answers; and (5) the public is neither thought to nor expected to have good ideas.[17]

Such top-down structure negates lasting citizen involvement in the planning process, reduces emotional investment in the product, and results in plans that have limited meaning or strength of focus to guide a community through change. Without a shared community vision for a shared, sustainable future, which can only come from the bottom up, plans either serve a politically invested few or are stored and go unused, unimplemented.

Building trust with and among the citizenry is an essential ingredient in any planning process, which means transparency and frequent communication. A high-profile, highly visible, open, and transparent process maintains confidence within the community. To achieve such confidence, however, one must make absolutely clear what the various stages of the planning process are; what delineates the transition from stage to stage; who makes decisions and how; when meetings will occur, as well as their purpose; and, of major importance, what will happen with the information provided by the public. Press releases, an interactive website, radio and TV public service announcements, and newsletters all work to keep the public educated, informed, enthused, and involved.

Preparing for a meeting starts well before the meeting occurs. The following guidelines should help in framing productive and interesting meetings:

1. *Ensure objectives are clear.* This is done through collaboration and lead to clarity about the desired outcomes of each meeting.
2. *Characterize the situation and identify interested parties.* Design a systematic method for identifying interested parties critical to the meeting's objectives and thus understand the participants' needs, interests, and expectations to ensure that information can be factored into the collaborative process design.
3. *Design and guide a highly credible process.* Ensure that participants have—and trust—the information necessary to enlist their full engagement in the process and in order to make decisions.
4. *Facilitate engaging and productive meetings.* Ensure that your virtual or in-person meetings are well structured and well run.
5. *Drive the progress with cogent documentation.* Capture all perspectives, accurately reflect common ground when present, carefully identify remaining issues and areas of difference, and create ownership among all parties.
6. *Sweat the details.* Meetings require seamless planning and logistics; appropriate technology; a thorough understanding of issues; efficient

and effective meeting agendas and materials; and thoughtful meeting summaries. Documents that track agreements, commitments, action items, and remaining issues are crucial. Ensure that participants are never surprised.

The conscious use of language is critical. It is important for planners and other technical professionals involved in the planning process to eliminate the use of professional jargon, technical terms, and agency lingo when interacting with the public. A "breakdown" in communication can, and most often does, have drastic impacts, stifling the willingness or courage of participants to offer creative ideas. It is the responsibility of the planners to learn the language of the public—not vice versa.

Ground rules that are approved by participants are essential. Such rules could, for example, include the following, adapted from many resources and years of framing ground rules for meetings by the authors:

1. Listen without interrupting when someone is speaking.
2. Participate in both listening and responding.
3. Be courteous at all times. Be open to other viewpoints; critique ideas—*not* people.
4. Be honest: participate openly, share what you think; do not withhold ideas.
5. Be positive; possibility driven.
6. Avoid side conversations.
7. Everything in the meeting stays in the meeting. Honor confidentiality.
8. Regard each person's input as important.
9. Reset your thinking from "me" to "we."
10. Share your ideas as "a" way instead of "the" way.
11. Stay focused on the topic—limit or, better yet, avoid war stories.
12. Be mindful of time and make your comments as succinct as possible.
13. If you have an issue, concern, complaint, or problem, please bring it up, but only with a proposed solution.

Another aspect of keeping communication free flowing is the recognition of the diversity and multidimensionality of the participants. In the context of the planning process, it is important to remember that emotional data (personal feelings) are as valid and legitimate as factual data. Feelings, whether translated into data by those who can afford the time, energy, or dollars to do so or those simply expressed openly and willingly, translate into political clout and must be respected equally. Feelings are a critical fuel for a planning process and must be acknowledged and used if the planning process is to succeed over the long term.

Sometimes, participants cannot get beyond their feelings. This can stymie their ability to communicate. In cases such as this, a facilitator might ask a participant for clarification; that is, the facilitator can paraphrase what he or she heard and then ask if it is correct. In addition, the person can be encouraged to offer a possible solution to the perceived problem that gives rise to the strong feelings, which sometimes helps to bring out possible solutions.

Above all, involve the public in the planning process from the very beginning. A well-designed planning process keeps citizens involved in civic life long after the plan is created. This is but saying that the process becomes more important in some respects than the plan itself because continued positive, productive engagement by the citizens of a community in creating their future and the future possibilities for their children and their children's children is a true characteristic of sustainable community development within the framework of social-environmental sustainability.

Open Space Planning, an Alternative Process

An interesting method for enhancing the flow of communication in the context of community planning is through "Open Space Technology." In the late 1990s, I (Jane) engaged in such a planning process. This innovative technique "taps into the dynamics of self-management, collective leadership, self-directed work teams, and genuine community. It works best in times of accelerated change with high levels of complexity, diversity, potential conflict, and time pressures."[18] Basic premises of Open Space are that whoever is there is who needs to be there; whatever is discussed is what needs to be discussed; and when it is over, it is over.

The purpose of the Open Space planning session I attended was to envision various futures for the Sigurd Olson Environmental Institute of Northland College in Ashland, Wisconsin, where I (Jane) was employed. A diverse group of participants attended this day-long event—members of the faculty, administration, and staff of Northland College; members of the board of trustees and the institute advisory board; students; and people from the community of Ashland.

We gathered in a church, which had a number of rooms for our use, one for the larger gathering and others for breakout sessions. After a brief introduction addressing the purpose of the gathering, in this case to determine how best to perpetuate the legacy of Sigurd Olson, anyone who wished could enter the center of the circle and write on a large sheet of paper a topic he or she would like to discuss. After all who wished to do so created topic sheets, the sheets were posted around the room. At that point, anyone who wished to join in a discussion of a particular topic signed his or her name on the topic sheet. A participant could, of course, sign up for more than one topic. Time periods for discussion were added to each sheet.

In one such Open Space gathering, described to me by the organizational development consultant guiding this process, a secretary of a large corporation entered the circle and created a topic sheet, posing the question: "What's really going on in this place?" This question drew the largest discussion group and resulted in significant changes in the organization.

Open Space Technology provides the format for gatherings without agendas; all who wish to join in may do so. The results that emerge from the hearts and minds of the participants in an atmosphere of equality and openness are far reaching in terms of follow-up actions. The form is nonlinear and self-designed. Human creativity is the core function and the driving force of this process. The results of the session I attended continue to penetrate the decision-making process of the institute and have a kind of unique durability.

Step by Step: A Suggested Process for Developing a Comprehensive Plan

A community's comprehensive plan needs to capture the essence of what people living in a place would like to sustain. In other words, the plan not only must be shaped by community values but also must be a strong vessel for those values. In developing a comprehensive plan tied to community values, creativity, imagination, and innovation are useful tools to access. How do we help communities "think outside of a self-imposed box"? How can we help communities access their creative, imaginative, and innovative selves to maximize the usefulness, meaning, and community-wide importance of the plans they create?

Setting the Stage for Planning

Setting the stage properly for planning is critical to the overall process. It represents an important part of leadership: knowing your audience, knowing their sense of "place," and knowing their history.

Setting the stage for planning involves gathering all available information about a community that is reasonable within time/cost constraints. Materials to gather include recent census data, any long-term studies of community demographics, previous comprehensive plans or planning studies, economic studies, maps, city budgets for the past 5 years, any technical or scientific studies relative to the community's health and well-being, and any information that describes existing conditions. A view of a community's history is usually available in concise form, either through the local historical society, in public documents, or even in previous comprehensive plans.

During this phase of planning, identify, with the help of the community's elected and hired officials, people in the community who are important to interview; these people need to represent a cross section of the community. Interviewing important people in the community is a multipurpose activity that helps the planner-facilitator obtain what we call an "issue-actors map" of the community. Through one-on-one interviews, we can gather information about both surface and subsurface issues, concerns, ideas for change, and the existence of others in the community who might be helpful in providing a perspective and deeper knowledge of the community.

Simultaneously, maps not already available need to be prepared. A geographic information system (GIS) makes possible the mapping of any type of data available. Maps to gather include a community's zoning map; land use map; soils map; wetlands-floodplains map; maps of other environmental constraint (fault lines, for example); demographics map; functional classification map (streets and their physical conditions); and so on. All the gathered information can be used as reference materials during the planning process, as well as for the development of and inclusion in the planning document itself.

Other land use planning technologies are surfacing in full recognition of our limited understanding of the complex interactions of socioeconomic and environmental systems and the accompanying need to make smarter decisions regarding land use. The Land Use Evolution and Impact Assessment model is a comprehensive urban planning support system designed for decision making at a regional scale. Such a tool could be complementary for local, municipal-based planning.[19]

The Land Use Evolution and Impact Assessment model incorporates ecological, geographic, and environmental theories into a single hierarchical framework, yet is open and flexible. The model enables subsystems modeling in conjunction with a graphical modeling tool called the Structural Thinking Experimental Learning Laboratory with Animation, a software program that enables the user to dynamically visualize and communicate how complex systems and ideas work. It is a tool based on systems thinking and is frequently used in courses that teach systems thinking. A land use decision support system for the metropolitan region of St. Louis, Missouri, has been developed using the Land Use Evolution and Impact Assessment model approach.

All of the activities associated with the planning process must be made public through press releases, client/community website announcements, and so on. In addition, educating the citizenry by making the information easily available to the general public throughout the planning period will help to ensure or encourage people to remain actively involved in their community beyond the development and adoption of the plan.

In addition to announcing meetings and issuing interim progress reports, keeping an interactive website current and/or creating a newsletter can be effective in keeping the community engaged in the planning process. By visiting with the editor or manager of the local newspaper, it might be possible to have an ongoing column or display piece that, in addition to keeping the community abreast of the planning process, defines planning jargon or contains inspirational quotations that affirm the need for or meaning of planning for one's community as a benefit to all its members—present and future.

Although we discussed the importance and process of attracting a diverse group of citizens to public meetings previously in the book, it is important to highlight a few of the more critical concepts. To begin, it is vital during the early phases of the planning process to perform a community survey, designed with and aided by the citizen planning advisory team. Such a survey would help create the information base necessary for crafting a meaningful plan. There are many models for these types of surveys and software for manipulating available data.

Also during the early stages of the planning process, open meetings with the community in which citizens feel safe to express their ideas and feelings are essential to creating the plan for the community's future. These meetings may raise such questions as the following: When have we attracted enough people? Should there be a percentage-of-the-community threshold required for moving forward? Or, should demographic representation be the determining factor?

Another way of obtaining input is through a community charette. The word *charette* refers to any collaborative effort in which a group of designers drafts a solution to a design problem. For example, a community-wide, weekend-long charette was held in Bayfield, Wisconsin, to help launch its comprehensive planning process. In 2001, I (Jane) had been hired to lead this process.[20] Bayfield, a highly attractive, historic fishing town on the shores of Lake Superior, has a small year-round population (800), which soars in the summer as tourists flood the town from June through September, bringing a combination of economic, environmental, social, and cultural impacts with them.

The weekend event to launch the planning process began on Friday night with an inspirational message from Bayfield's long-standing and charismatic mayor regarding the community's future. The gathering place was a historic waterfront community center, formerly a herring shed at the turn of the last century.

Over the course of the next 2 days, a charette was held for the community to participate in designing their future. Planning activities for all ages were offered, along with food, beverages, live music, and planning resources, such as maps and public documents relating to the planning process. Children not yet old enough to write were asked in taped interviews what they liked about Bayfield and what new things they would like to see. A mature 9-year-old facilitated the activities for children, which included murals and writing letters to tell friends what Bayfield is like in an imagined future year of 2015.

For each element of the comprehensive plan, a table was available, hosted by a member of the planning team who was stationed there to gather input. At these stations, participants could offer their ideas regarding the specified element (land use, transportation, historic preservation, and so on) and what they thought was and was not working and what sorts of changes they would like to see. Some 300 people attended over the course of the weekend or about 37 percent of the population. The contributions received from the community were invaluable and helped inform a comprehensive plan tied to the principles and concepts of sustainability.

In 2010, Bayfield went further by developing a sustainability plan[21] in conjunction with the comprehensive plan based on The Natural Step, which the city had adopted in 2006, making the city an ecomunicipality.[22] Bayfield went on to receive countless awards and recognition, including being named the greenest community in the state in 2009 by then-governor Jim Doyle and, in 2010, recognition by the Global Management Education Center for being one of 38 communities around the globe that are leading the way in sustainability.

Step 1: Mining for Meaning—Obtaining Community Values

In developing a comprehensive plan, identification of community values is an essential first step. Value is defined as "something (a principle or quality) intrinsically valuable or desirable."[23] Finding out what the citizens care about is vital to the quality of a community's future and frequently falls into categories. A suggested three-part categorization of values is social, ecological, and economic. These values can serve as the present, nonnegotiable parameters or criteria for organizing the elements of the comprehensive plan and thus guide the spirit and content of the plan.

Community values can also be understood as those things that connect a person to a "place," and when retained, continue to feed and nurture that sense of place in and of a community. Recognition of and identification with community values also set the stage for community involvement in the creation of policies, laws, and regulations that are both meaningful and functional.

That personal values play a critical role in community is amplified by John Friedman in *Retracking America: A Theory of Transactive Planning*, which emphasizes mutual learning and conversation, where the "scientific knowledge of the planning expert is joined with the deeply personal, experiential knowledge of the client."[24] In Friedman's model, the joining of two distinct modes of knowing will guide community interaction. This interaction requires mutual self-disclosure of values, but if this kind of disclosure is missing, "messages may be exchanged, but the relevant meanings are not effectively communicated."[25]

The role of human values in the planning process is further explained by Ralph Keeny's *Value-Focused Thinking: A Path to Creative Decision Making*.[26] Values are described through the individual and group process of discovering the thinking or reasoning behind desired areas, especially where the need for change is perceived. Values surface through questions addressing desired outcomes and why a specific outcome is important.

> For most public problems, values, rather than facts, are the aspects of the problem about which many members of society will have knowledgeable viewpoints. Discussion of the details of the consequences of various alternatives often depends on technical and complex concepts from various professional fields. Hence, without discussion of values, many people are excluded from participation and others are limited to minor contributions. (p. 23)[27]

Step 2: Creating a Community Vision

When citizens take the lead in designing their future, there is an emotional investment in the outcome accompanied by a firmer commitment to its implementation. Further, when the process allows and encourages all participants to share their deepest feelings about their communities, a vision is strengthened and the heart of community revealed. Noted psychologist Carl Jung claimed that, "Your visions will become clear only when you can look into your own heart. Who looks outside, dreams; who looks inside, awakes."[28] On the community level, perhaps this can be construed to mean "that a community's vision will become clear only when citizens look into their community heart. A community that looks outside beyond their own center cannot be true to their uniqueness and possibilities. Any community that looks inside for its own potential, awakes."[29]

Creating a community vision can take many forms. I (Jane) once adapted a technique suggested by Geoffrey Ball, of Geoff Ball and Associates,[30] while working with the city of Ashland, Wisconsin, in the creation of a their comprehensive plan. I had recently moved from Monterey Bay on the coast of California to Ashland, a city of 8,000 on the shore of Lake Superior. The cultural, climatic, and demographic shift was significant, and I welcomed the challenge of being a planner in a place so unlike the place from whence I had come. The 30-member citizen planning group, appointed by the mayor to develop a comprehensive plan, was extremely diverse, from retired ore dock and railroad workers to hospital and college administrators, owners of small and large businesses, political activists, and members of community at large.

Prior to doing this work, I had visited as many of the "main actors" in the community as possible—those acknowledged leaders, as well as long-time community activists, and then those who had just been around for a long time and were well known in Ashland. My purpose was not only to get my bearings and learn as much about the community as possible before venturing forth in guiding the development of a comprehensive plan but also to disarm any fears someone in the community might have about me. After all, I was indeed an "outsider" in a part of a world where one remains somewhat suspect until living in the community for at least 20 years.

Here I was, fresh from California; driving a Japanese car, which stood out in a city hall parking lot dominated by General Motors; a female, in a mostly male-led town; and recently hired to guide them in community planning. I felt there was significant potential for community members to be skeptical since it seemed that many older people in Ashland had either not traveled or traveled very little and therefore had rather limited views of the potential for change in their town. In addition, many seemed unaware of the forces from outside of their community that might cause rapid, unwanted change in Ashland. Finally, many remembered the industrial past and the boomtown that Ashland once was, with an abundance of jobs and the activity, prosperity, and national prominence. This was a past that for some merited resurrecting.

With this in mind, my job, as I saw it, was to establish trust and to reassure those I visited that I was not bringing any particular template to the planning process, that I could not help but be influenced by my many years as a planner in a very different place, but that I would guarantee them an open process in which all points of view were welcome.

After an initial orientation of the planning task force to the job at hand, we undertook a visioning process, which, by the way, was not preceded by a process of soliciting values. My understanding of the process did not include that step at the time.

My introduction to the visioning exercise was to help them imagine they had been to a beautiful exotic place (I think I said Tahiti) where they had lived very comfortably. They had been in touch with a number of friends and relatives back home, who said that some wonderful and significant changes had taken place in Ashland, but provided no details. They were left only with their imaginations to determine what those changes actually were. At this point, I asked members of the task force to complete the story by jotting down their ideas of what they would have liked those changes to be and to elaborate further on how they would like their community to change.

Next, individuals were asked to form smaller groups, share their personal "visions," and form a collective vision to bring back to the larger group. Ball suggests that each group present a symbol to accompany its vision because symbols often reach beyond words. A list of at least five to seven suggestions for inclusion in a shared vision was requested as part of each group report to the whole gathering. This sharing of ideas is an important function in that it helps participants become more aware of the amount of commonality that exists with the group as a whole—a magic moment in which perspectives change about what is possible.

Following the magic moment, one member of each group is chosen to be part of a subcommittee to summarize all the data and develop a symbol for the vision. With data summarized and a symbol in hand, a meeting is called at which the summary is reviewed by the full group; the group adds, modifies, or revises elements or parts of the vision. Once consensus is reached, the group signs the vision statement as a record of its acceptance.[31]

Aided in large part by community input obtained during a weekend-long charette, described above, a vision for Bayfield, Wisconsin was created. Using this input, the planning team prepared a draft vision and then took it back to the community at large with an event as celebratory and open as the charette. Each member of the team was asked to write their own vision stating his or her "preferred future" in each of the community areas now identified as elements of the comprehensive plan. The only rule was to write in the present tense for the year 2020. Some questions were offered for consideration while framing their vision:

1. What is it you value most about Bayfield?
2. What needs to change, and what does the change look like?
3. What threats, if any, do you see to your desired changes or to the current ideal state?
4. What opportunities exist that you feel merit attention, time, and resources?

Because all wrote their visions via word processing and spreadsheet, it was relatively easy to merge the statements, discover the discrepancies, and produce one vision, highlighting, of course, the areas of disagreement. The final vision statement was then brought to the community, where it received broad approval.

Step 3: Preparing Elements of the Plan

We propose that in developing a comprehensive plan, each element contain the following:

1. A statement of perceived necessities and values stated broadly enough to encompass the requirements for integration of human and nonhuman living systems
2. A description of the current state of the particular component of the community system (composition, structures, functions, and services)
3. An analysis of what is working and what is not and why
4. What external conditions offer opportunity or pose threats
5. Goals, objectives, policies, and outcomes

In addition, the results of this planning process should include subplans for implementation, monitoring, and periodically revisiting the vision statement and parts 1 through 4. Thus, the planning we are recommending, as an ongoing process, is really a way of thinking. The plan is never perfect or complete and thus is a "community journey," if you will.

What distinguishes the format we are recommending from the conventional format of most comprehensive plans is accounting for the structure, function, goods/products, and services of both the human and nonhuman components of the system in designing a desired outcome for human future.

Typically, a comprehensive plan has an element called "environmental quality" that primarily defines constraints to development. In the proposed comprehensive planning format, ecological considerations—including constraints based on the inviolable, biophysical principles laid out previously in this book—are built into each element of the plan. Further, this approach places strong emphasis on community values, both broadly speaking and in association with each element of the comprehensive plan.

A Top-Down Planning Process: Lessons from the Field

The ideal discussed of a truly collaborative process is not always allowed, however, as you will see in an example of top-down planning from my (Chris's) hometown of Corvallis, Oregon.

As you read the account, keep in mind that when we attempt to redesign nature, increasingly through the technology of bioengineering, to fit our desires of a risk-free system, we build in two hazards. One, we build rigidity into the system that it will one day reject—to our peril; two, we design the system to mirror our own human limitations, including our limited understanding of the biophysical relationships that govern nature.

This drive to remove all risk essentially sets the stage for greater risk, as Aldo Leopold once said, "Too much safety seems to yield only danger in the long run."[32] We are not, after all, masters of the tools and technology that we invent. They affect our lives—and those of our children and their children and their children's children—in ways we most often do not understand and, in the long term, cannot control.

Bioengineering versus Natural Processes

"The dream that technology serves for us," writes John Gray, professor of European thought at the London School of Economics, "is a dream of complete control. It's the dream that we can cease to be mortal, earth-bound creatures subject to fate and chance."[33] Gray goes on to say that the use of technology to remake the world according to our human desires is not so much a product of science but rather of magic and is the fantasy by which industrialized nations have been ruled for much of the twentieth century—and so far the twenty-first century. What, specifically, you might ask, is this fantasy? It is the notion of progress without risk, instability, and uncertainty.

"When we look to technology to deliver us from accidents, from choice, and even from mortality, we are asking from it something it cannot give: a deliverance from the conditions that make us human." What we once looked for in political ideologies, and before that sought in religion, we are today seeking from technology, contends Gray—"salvation from ourselves."[34]

In this sense, says Gray, "The world today is a vast, unsupervised laboratory in which a multitude of experiments are simultaneously under way."

This premise can be applied to the so-called restoration of the riverbank that borders the eastern side of downtown Corvallis, which took place in 1999.[35] Downtown Corvallis is situated on the west bank of the Willamette River. The waterfront portion of the downtown area was to be renovated and 2,750 feet (69.9 meters) of the riverbank restored to greater ecological health for a cost of $4 million, which is the issue of this discussion.

The city council opted to convene a meeting after a proposed change in the plans to "stabilize" the riverbank prompted outrage among Corvallis residents. The proposed change was the cutting of 400 trees. The riverbank restoration work was part of a massive project to renovate the riverfront by constructing a new Riverfront Commemorative Park and to meet federal requirements to stop the overflow of sewage into the Willamette River that occurs during heavy rains.

The city council originally approved a plan that sought to save as many of the existing trees along the upper part of the riverbank as possible, but the consultants hired by the city said their data showed that saving those trees would pose a threat to the stability of the bank. Some residents, including a team of 21 scientists from Oregon State University, opposed the consultants' recommendations.[36]

The team of consultants included Gordon Nicholson, riverfront project manager for the engineering firm CH2M Hill, and Robbin Sotir, an expert in soil bioengineering from Marietta, Georgia. According to the consultants, a computer analysis of soil tests showed that "deep weaknesses" and "significant threats of slides" occurred in the riverbank. Problems contributing to the crumbling riverbank were said to include seepage, leaning and top-heavy trees, cracking along the upper bank, and debris of metal and concrete in the upper bank. "These [taken] all together," asserted Nicholson, "are saying this is a bank with symptoms of stress."

In short, and grossly simplified, the bioengineering approach to stabilizing the riverbank included (1) removing about 400 live trees; (2) cutting the riverbank into a series of terraces and reinforcing it with packs of soil wrapped in a geotextile material; (3) piling layers of boulders (called riprap) on the bank to protect it from being washed away by flood waters; and (4) planting new trees.

"I think everyone would like a low or no-risk option," Nicholson said, "and those were our marching orders." Such thinking implies a disaster mentality that conjures the worst possible case scenario, one that requires changing the world in an attempt to prevent it from happening.

Nicholson went on to justify the bioengineering approach to stabilize the riverbank by indicating that the number of trees, including tall trees, would be increased over the number of trees that occupied the bank during the 1970s. Sotir added that the same bioengineering technology being proposed for the riverbank in Corvallis has been used in 30 other countries and along the Mississippi River.[37]

Saying that the same bioengineering technology has been used elsewhere in no way indicates whether the technology was appropriate; whether it worked; how well it worked; how long it has been in place; whether it has been evaluated; or what ecological problems, if any, arose and/or were created. Bioengineering, at least in my (Chris's) experience over many years, focuses on the engineering in response to a perceived disaster with precious little understanding of the ecosystem involved or its processes. Bioengineering is too often viewed and practiced as a quick fix that puts engineered rigidity, which is presumed to build predictability or certainty, into a fluid problem governed by ecological processes, which often spawns disastrous results—such as the predictable, rigid levees placed along the banks to control the fluid, unpredictable, and uncontrollable Mississippi River.[38]

On the other side of the coin, scientists from the Department of Fisheries and Wildlife at Oregon State University in Corvallis disputed the consultants' conclusion, which, as it turns out, was never subjected to a disinterested third-party review. The scientists, particularly Boone Kauffman, a riparian plant ecologist, and Stan Gregory, a stream ecologist, carried the argument that the consultants overstated the threat of the bank crumbling into the river. Their counterpoints to the consultants' report are as follows:

Riverbank Instability and Its Risk of Failure

1. The Corvallis riverfront is composed of geologic materials associated with Missoula terraces formed more than 10,000 years ago. The material forming the Missoula terraces is much more resistant to erosion than are alluvial gravels, which are characteristic of other locations. Alluvial gravels are those that have been deposited by flowing water, such as a river.
2. The Corvallis riverfront is among the most stable sections of the Willamette River from Corvallis to Eugene, a town about 50 miles (81 kilometers) to the south. It has changed very little in the last 150 years, according to historical maps and historical survey data.
3. The Corvallis riverfront has experience 20 floods since 1860 that were larger than the one of 1996 and has survived all of them with no evidence that the river has cut into its bank more than a few tens of feet (meters).
4. The consultants' estimation that the riverbank has been eroding at the rate of 1 foot (0.3 meter) every 3 years is a gross overstatement. A cursory review of the riverfront will confirm that the majority of the bank is well vegetated. If the rates of erosion were as stated by the consultants, there would be few trees left on the riverbank, whereas now there are many.
5. There is no evidence of tension cracks to indicate that major planes of failure extend back from the riverbank, just as there is no evidence

or slumps or bank erosion that extend back more than 10 to 25 feet (3 to 7.6 meters) from the riverbank over the last 150 years.

6. Cracks along the riverfront bike path are similar in structure and frequency to those that occur throughout the bike paths of Corvallis, even those well away from creeks and rivers.
7. Only two slumps are clearly related to the 1996 flood, which was severe. Others, which have occurred historically, were likely exacerbated by problems associated with drainage of the city's storm water.
8. It is difficult to assess or predict the risks of bank failure without analyzing the multitude of natural processes that govern the floodplains of the Willamette River in both time and space. It is therefore necessary to take a larger view of the processes that affect the riverbank rather than a microstudy of those characteristics that affect only the site in question.
9. There is no evidence to support the belief that the proposed bioengineered geotextile structures will perform better than, or even as well as, the existing bank structure and accompanying vegetation, which is already well established. Commitment to this extensively engineered and unproven design may well be ecologically unsound and thus an unwise financial risk for the community, as the bioengineered design requires stripping most of the existing natural structure and vegetation from the already long-stable riverbank in favor of a largely artificial approach to preventing a potential and unlikely erosional disaster somewhere in the unforeseeable future.[39]

The Riverfront Forest

1. Tree composition of the riverfront is today a largely uneven-aged mosaic of healthy indigenous riparian hardwood trees. Not only are the majority of the trees young and vigorous but also the arborist contracted by the consultants to examine the trees found that 80 percent of them were in the highest class of health and vigor.
2. The overall good health of the trees along the riverbank appears to be the norm for the forest that borders the Willamette River. The current richness of species is high, with over 40 different species of plants identified, and the structural diversity of the riverfront is continuing to increase and will do so for decades to come—provided it is not stripped in the name of bioengineering.
3. Given the present health and vigor of the trees, there is no accelerated risk of trees falling into the river when compared to any other riparian forest along the Willamette River of similar age and composition.
4. Scientists throughout the world recognize and acknowledge the role of vegetation in providing soil stability because plant roots bind soils

together. The root mass of the riparian zone is equal to or greater than the mass of trees that are seen aboveground.

5. A survey conducted by the scientists showed that the natural falling of trees not only appears to be an extremely rare event along the town's riverfront but also is not associated with bank failure.
6. The density of tree stems dissipates the erosive energy of high water during floods; in addition, the increased number of trees over the last 50 years indicates that the stand is becoming ever more stable and increasing in its ability to dissipate the energy of flood waters.
7. The pistol-butt growth pattern of trees along the riverbank may be caused by conditions that are unrelated to bank instability, such as stems that sprout laterally from the butt. Straight stems above the pistol butts, on the other hand, indicate long periods of bank stability.[40]

The Outcome of the Proposed Project Is Uncertain

1. Most projects similar to that proposed by the consultants have been conducted where erosion is severe and banks are largely unvegetated, which is not the case for the Corvallis riverfront.
2. If the consultants' plan is implemented, the current structure of the forest not only will be destroyed but also will take many decades to regain its prominence. The use of geotextile fabric under the soil will result in a soil depth of 2 feet (0.6 meter), which will limit the establishment of indigenous trees and will depress the growth of both trees and shrubs.
3. Projects, such as that proposed by the consultants, have not been conducted in Oregon's climate, where patterns of winter precipitation and summer drought are typical. Drainage within the project may prevent the survival of planted vegetation due to water stress during the summer, a common problem for riparian plantings, even in the best of soils.
4. The trees at the base of the project covered by riprap will not likely survive as depicted in the drawings presented by the consultants. The placement of large boulders is damaging to the trunks of existing trees, subjecting them to disease and rot. In addition, the riprap surrounding the trees will act like a choke collar by restricting the ability of the trees to grow in diameter.
5. Finally, based on my (Chris's) own experience, the large rock riprap that is being used to form a layer of "protection" over the riverbank will most likely become an ideal place in summer for the breeding of Norway rats, something already plentiful in the city of Corvallis.[41]

The Oregon State University scientists recommended integrating the existing conditions of the healthy riparian forest into the plan for the riverfront to maintain the current and vital ecological functions and the aesthetic qualities. Their basic recommendation was simple: *Work with nature*—not against nature.

Gordon Grant, a research scientist and expert on rivers and how they flow, reiterated this admonition by saying there is little evidence that the downtown reach of the Willamette riverbank is unstable. He further contended that the CH2M Hill consultants hired by the city did not clearly establish a problem with the riverbank before moving ahead with plans to "stabilize" it.

"At this stage, I don't think the documents support the claim of a bank stability problem," Grant said. "If this is what the ... [need to stabilize the riverbank] is pinned to, ... [it's] on thin air. Everything they [the consultants] did was to rectify a bank stabilization problem that was never documented." Grant feels the consultants represent a "rush to judgment," because, he muses, "if you have a hammer, everything looks like a nail."[42]

Dick Gamble, a civil engineer with the U.S. Army Corps of Engineers, was familiar with the riverbank in question and agrees. While the corps had not studied that particular stretch of riverbank, said Gamble, the corps had, as part of its job of controlling floods and maintaining waterways, studied the banks just north and just south of the downtown area, which have needed little attention, Gamble said, because "historically, that reach of the river has been pretty stable."

Dick Bell, also a civil engineer living in Corvallis, retired from Oregon State University, where he taught for 30 years. Bell is a noted expert in the mechanics of soil, foundation engineering, and the use of geotextiles, which the consultants recommended using to stabilize the riverbank. "I haven't been aware of any problems," said Bell. "You don't see a lot of trees falling down," Bell continues, and "the fact that it has been stable—common sense says it will remain stable."

Dr. Bob Beschta, a professor of forest hydrology with Oregon State University, said the consultants mistakenly blamed the river for eroding the bank and used damage inflicted by the 1996 flood as an example of what could be expected in the future when viewing the riverbank with a disaster mentality. Instead of the river being at fault, said Beschta, the intensity of the rainfall and the speed with which it moved along pavement were to blame for the 1996 landslide. This means, Beschta continued, that the city must find a way of collecting the storm water runoff and lessen its effect on the riverbank.

Gordon Nicholson, of CH2M Hill, on the other hand, said they, the consultants, "were to come up with solutions to protect the city's investment in the riverfront" and that *"doing nothing was not seen as a solution"* [emphasis added]. But, many citizens wondered, myself included, where the problem was. Has anyone considered, one might therefore ask, as I did at the October 18, 1999 meeting, what is right with the riverbank as it is, from an engineering point of view. No reply was forthcoming.

Perhaps the problem has more to do with minds predisposed to finding a perceived "no-risk" way of protecting economic investments within the framework of a disaster mentality than it does deriving an ecological understanding of riverbank processes and the real potential of soil erosion. In other words, the mayor, city council, and consultants failed to differentiate between *possibility* and *probability*, which are two vastly different dynamics.[43]

Listening—Really Listening—to the Citizens

Listening, really listening, to the citizens requires an open mind, which has not been the case in Corvallis. "The riverbank plan has exploded and debris is falling around our ears," wrote Corvallis columnist Wendy Madar. She went on to say, "As a regular at Corvallis public meetings for many years, I've had the uneasy sense that public comment is not always as welcome in fact as it is in name." Madar said her unease was crystallized during a recent meeting of the city council in which a councilor told a member of the community that he had forfeited his chance to contribute to the riverfront plan because he had failed to attend the original planning meeting. "It's too late now. I'm sorry!" the councilor said in startlingly harsh tones.[44]

The city staff also used the phrase "eleventh-hour stakeholders" to characterize citizens who did not speak out earlier about the riverfront plan. The implication is either that these citizens failed to contribute to earlier meetings and thus had no right to speak now, or they were coming forward to "get what they didn't get the first time around."[45]

Although city leaders have repeatedly stated that process leading to the riverfront plan was both long and exhaustive and, therefore, must represent a consensus of the citizenry, many members of the community tell a different story. They say that while they did indeed attend the meetings and presented their concerns, the results did not reflect their testimony.

At an October 18, 1999, public meeting on the riverfront project, I (Chris) got the distinct feeling, especially from the mayor, that the public was a nuisance to be endured, albeit grudgingly. One of my neighbors came to me the next day and told me that he and his wife had both attended the meeting and also had experienced the exact same feeling. In fact, of the 42 people who spoke at the October 18 meeting, most opposed the consultant's plan. I was among them.

This discrepancy points to a real and continuing problem with communication between citizens and city officials, particularly when new or unwelcome issues are raised. I (Chris) can attest to this from years of personal experience. Consider the following example of miscommunication: A botanist, testifying before the city council, said that the work of putting the layers of rock (riprap) along the riverbank *should be stopped* because it was not needed given that the bank has been historically stable and that the rock was destroying the healthy riparian plant community. When he finished, an official thanked him for speaking out *in favor* of the city's plan for the placement of underwater structure for fish habitat.

Dr. Boone Kauffman said that people where looking for and expecting restoration of the riverbank and improving the beauty of the riverfront park when they voted. "People voted for restoration and now we're getting stabilization."

"I keep listening for a voice in the community arguing for an engineering approach [to the riverbank restoration]," said Dr. Stan Gregory, "and I haven't heard it." Well, there may have been one; Tom Peterson wrote the following in the local newspaper:

> I write this after having walked the length of the riverbank from Harrison Boulevard to Washington Avenue. The view to the east is little different than walking through any undeveloped forest area. Yet, I know that there is a beautiful river on the other side of the trees on the riverbank that begs to be viewed.
>
> The current view of the river is occluded with trees and brambles. What a mess. It seems to me that hacking down the 400 trees and doing what's necessary to eliminate erosion of the riverbank will afford us a greater view of the river and give us an opportunity to landscape the riverbank so that it is an asset to the community rather [than] the jungle that it currently is. For those that prefer the view of the trees, there will still be plenty on the east bank of the river.
>
> Start the chain saws.[46]

Although miscommunication may not be a deliberate move to limit public debate on an issue, it often has that effect. Saying that a process is public is therefore not enough; citizens not only must be allowed to speak freely and safely but also must be listened to with respect and heard with patience and accuracy. "There is no such thing as an eleventh-hour stakeholder," says Madar, and there must be time in a democratic process to listen to informed citizens.[47]

Or Only Pretending to Really Listen

As it turned out, stabilization of the riverbank was not the central issue, but rather a two-way street with parking that would encroach right up to the functional edge of the riverbank, where there supposedly was to be a park with open space for people to enjoy walking and biking along the riverbank itself, which means there is only room for the existing one-way street without parking. But, apparently, the city council had made its mind up months ago to please the businesspeople who wanted the two-way street and its attendant parking spaces to be the nonnegotiable centerpiece of the entire riverfront plan. And, it turned out the city council had already voted on the various elements of the plan for the riverfront park and bank stabilization they wanted implemented.

The citizens, however, were not openly informed about this part of the plan in the beginning, and even when the two-way street became common knowledge, raising the ire of many citizens, as evidenced by comments in public meetings and many letters to the editor in the local paper vehemently opposing the two-way street, the city council simply ignored the comments.[48]

In addition, the scientists from Oregon State University, who were appointed by the city council to a peer-review team to check the ecological accuracy and accountability of the stabilization plan developed by CH2M Hill, expressed their dismay when the main point of their review was never discussed at the January 6, 2000, public meeting. The scientists' point was that much expensive bank stabilization could be avoided simply by moving the park and street improvements some distance back from the edge of the river, but that would mean giving up the two-way street in the narrowest part of the proposed renovation.[49]

As might be expected, the scientists were charged by the city council to be *neutral* in presenting their findings. Moreover, Dr. Bob Beschta said that "essentially, no comments were made" about the scientists' report by the city council at its January 6 meeting. Although the review team disbanded after handing its report to the city council, some members wanted to ensure that the public fully understood the implications of their report because, said Beschta, our "report was basically sort-circuited … [and] we wanted to keep this process public." They thus decided to hold a public forum, which took place on January 19.

Dr. Dave Bella, an Oregon State University professor of civil, construction, and environmental engineering, said that it was up to the city council if they wanted to spend the money in the narrow area to shore up the riverbank as planned. But, he continued, a second option, outlined in the report, did not even get a hearing. That option was to save the money slated for stabilization of the riverbank and move the park improvements away from river or eliminate them, which would leave more open space and cost less. "If you never look at this option," said Bella, "the cost gets hidden. To not have it [all the options in the report] laid out—that makes me uncomfortable because people are liable to say that they [the city council] had to do it [follow the heavily contested original plan] because of your report."

Dr. Stan Gregory said the Willamette River should drive the discussion, which means city officials should plan developments so they will not be affected by the river and thereby respect its dynamics, rather than continually attempting to subjugate the river. But, the river's dynamics remained ignored by the city council in favor of a two-way street. There was yet a more egregious problem concerning information about the riverbank, one internal to the city.

Eliminating Unwelcome Voices Within

"If I were to tell you," writes Corvallis columnist Wendy Madar, "that, nearly a year ago, [Greg Peterson] the engineer in charge of the city's riverfront sewer project raised many of the same concerns now being raised by the citizens [about the riverbank stabilization project] and soon afterward was fired, you'd very likely say, 'Not in Corvallis.'" But, that is exactly what happened, says Madar, as indicated by papers given to her by someone other than Greg Peterson, including letters in which Peterson, who had overseen hundreds of complex environmental projects in his 24 years with CH2M Hill prior to accepting his job with the city, raised his concerns. Although the city officials said Peterson's dismissal had nothing to do with the questions he raised, the fact that he raised them so early casts "a strange light on what we're now being told."

Peterson, who was the city's overall manager for the multiple sewer and riverbank projects, soon became uneasy with several aspects of the work on the riverbank, including inadequate information being given to the citizens and discrepancies between what voters had approved and the developing bioengineering design. Peterson wrote letters to Gordon Nicholson, who you will remember was riverfront project manager for the engineering firm CH2M Hill, city manager Jon Nelson, and others regarding the plan for the riverbank.

Peterson's questions are striking, writes Madar, in that they echo issues raised at recent public meetings, such as the necessity of removing vegetation and soil, the wisdom of destroying existing soil-holding root structure, the environmental soundness of the geotextile technology as proposed, the potential for bank failures during construction, and the thousands of loads of debris that would have to be trucked through downtown Corvallis.

Peterson had been scheduled to address the Downtown Corvallis Association on December 16, 1998. Joan Wessell, director of the Downtown Corvallis Association, told Madar: "We were delighted to work with him. He's an excellent communicator." The presentation never took place, however, because Peterson was fired at 4 p.m. on December 15.

Peterson told Madar that, "It was the most incredible thing that has ever happened to me." He said that city officials had told him not to raise difficult issues with business owners concerning construction along the riverbank, especially the specter of 50,000 dump truck loads of soil being moved through downtown Corvallis, "because it will confuse the situation." Peterson, on the other hand, saw communication with the citizenry as the most important part of his job.

Why, wonders Peterson still, did the city officials not consider a wide range of options or have a peer review conducted on Robbin Sotir's geogrid design before deciding what to do, but he speculated that the lure of a "risk-free" riverbank diverted the process away from the ecologically oriented approach that the community both voted for and expected. Sotir's design, says Peterson, ironically has significant risks associated with it, risks Peterson detailed in his letters.

Now that the citizens are raising the same question, says Madar, "[City] officials are telling us these are new concerns that surfaced as a result of recent soil tests."[50] How, one might wonder, can these be construed as new concerns when Peterson raised them a year ago?

A major point used by city officials in persuading voters to approve a $9.5 million bond for the riverfront restoration project was wrong, according to city records and to Gordon Nicholson, the lead consultant, who said that the rate of erosion cited in the city's documents and the reports by CH2M Hill was incorrect, which the city is blaming on CH2M Hill. Nevertheless, voters approved the bond to fund restoration of the riverbank and construct a riverfront park because city officials said that the riverbank was unstable, and that without several million dollars of engineering work, the park and nearby businesses, as well as power, sewer, and utility lines, could be threatened by the eroding bank.

But, this sorry saga does not end here. Those in charge of the city of Corvallis did not follow their own rules for competitive bidding when they awarded to CH2M Hill the contract to restore the riverbank, the most controversial part of the riverfront project. "The city decided CH2M Hill was the only company that could do the job—a $147,325 contract to develop preliminary designs for restoring the downtown riverbank," writes Aaron Corvin of the *Corvallis Gazette-Times*.

Although there is a "sole-source" exemption that will allow a city contract to be awarded to a company without inviting bids in a public process if the goods or services are available from only one company, the city's own rules required it to document why CH2M Hill was the only company that could do the job, which was not done, according to Nancy Brewer, finance director for the city. The city apparently awarded the contract for the initial restoration design to CH2M Hill in September 1998 and received a draft of the design in October of that same year—about a month before the voters approved the $9.5-million, 20-year bond measure to help fund the project, at which time the design not only failed to show how the trees would be saved along the riverbank but also stated that most of them were in poor health.

On top of all this, Gordon Nicholson, riverfront project manager for CH2M Hill, was quoted as saying: "I'd have to say the city's procedures appear to be consistent with the norm, realizing that there's a fair amount of variation from one agency to another." Be that as it may, the city officials of Corvallis lost credibility by hiring CH2M Hill to begin designing the riverfront without competitive bidding. Although the officials listed logical reasons for their actions, their decision remains open to question because they broke their own city rules when hiring CH2M Hill as a sole-source contractor without the required competitive bidding.[51]

Does this in any way sound familiar? It would not be surprising if it does since this kind of saga is replayed in city after city, where people use the symptomatic linearity of their thinking to try to control in a predictable, risk-free manner the often-unpredictable, systemic, cyclical processes that govern the biophysical circumstances of all living things—including us.

Summary

Chapter 7 introduces the importance of and ways of modeling the planning process after nature. Modeling land use planning after nature is a vital step in our human journey toward social-environmental sustainability. Chapter 8 examines the best means whereby to implement a comprehensive land use plan.

Endnotes

1. Parker Palmer. http://www.wisdomquotes.com/authors/parker-palmer/ (accessed February 27, 2013).
2. *The New Merriam-Webster Dictionary*. Merriam-Webster, Springfield, MA, 1989.
3. V. H. Dale, S. Brown, R. A. Haeuber, and others. "Ecological Principles and Guidelines for Managing the Use of the Land." *Ecological Applications* 10 (2000):639–670.
4. John Jackson. "Resources, Not Garbage: Municipal Solid Waste in Ontario." Prepared for the Environmental Agenda for Ontario Project [Canada]. March 1999. http://docs.google.com/a/peak.org/viewer?a=v&q=cache:-YwzjPi59BUJ:cielap.org/pdf/EnvAgenda_Resources.pdf+Resources,+Not+Garbage:+Municipal+Solid+Waste+In+Ontario.&hl=en&gl=us&pid=bl&srcid=ADGEESgPVOaXpXviqjL6UMaRFxIoeaeEgD2OGqwhH9PhZD8xW6Y0LpJt9f-galt3b2o5gvFClOzLU-mz8A6LViPlcAbWAsoyNWQtb2ShnMTdIhq_Iie233ZB_RD0TL78pTuZrArB1RE-7&sig=AHIEtbRKEiVJSZzXAyaTN570oJUKYrk4zQ (accessed March 2, 2013).
5. Chris Maser. *Ecological Diversity in Sustainable Development: The Vital and Forgotten Dimension*. Lewis, Boca Raton, FL, 1999.
6. Peter Senge. Keynote speech. Systems Thinking and Actions Conference. Atlanta, GA, November 5, 1999.
7. Alfred North Whitehead. http://www.wisdomquotes.com/quote/alfred-north-whitehead-3.html (accessed February 28, 2013).
8. Margaret Wheatley. Presentation to Bainbridge Graduate Institute/Organizations Systems Renewal, Seattle, WA, October 18, 2013.
9. John Lennon. http://www.thequotefactory.com/quote-by/john-lennon/there-are-no-problems-only-solutions/44662 (accessed February 28, 2013).
10. Chris Maser. *Principles and Concepts of Sustainable Community Development*. CRC Press, Boca Raton, FL, 1996.
11. David Orr's characteristics of ecological sustainability as referenced in Sim Van der Rym and Stuart Cowen. *Ecological Design* (10th anniversary edition). Island Press, Washington, DC, 2007.
12. Braden Allenby principle, as summarized by Jay Moynihan. "Sustainable Development." August 8, 2012. http://www.airwaterground.com/Sustainable Development-definition-jsm-v08092012.pdf (accessed February 28, 2013).

13. Donella Meadows. "Places to Intervene in a System." *Whole Earth* Winter (1997):78–84.
14. Warren Nelson. "Riding the Wind." 1985. http://www.warrennelson.com/biography.html (accessed March 2, 2013).
15. Chris Maser. *Vision and Leadership in Sustainable Development*. Lewis, Boca Raton, FL, 1998.
16. Ralph Waldo Emerson. http://www.brainyquote.com/quotes/authors/r/ralph_waldo_emerson.html (accessed March 2, 2013).
17. James L. Creighton. *Citizen Participation/Public Involvement Skills Workbook*. Synergy Consultation Services, Cupertino, CA, 1972.
18. Harrison H. Owen. *Open Space Technology: A User's Guide*. Berrett-Koehler, San Francisco, 2008.
19. The foregoing discussion is based on Zhanli Sun, Brian Deal, and Varkki George Pallathucheril. "The Land-Use Evolution and Impact Assessment Model: A Comprehensive Urban Planning Support System." *URISA Journal* February 8, 2005. http://www.urisa.org/publications/journal/articles/the_land_use_evolution (accessed March 13, 2013).
20. City of Bayfield. "Bayfield Comprehensive Plan." http://www.cityofbayfield.com/poster-plan.html (accessed March 13, 2013).
21. "City of Bayfield, WI. Sustainability Plan." July 30, 2012. http://www.lwm-info.org/vertical/sites/%7B92F7D640-E25A-4317–90AD-4976378A8F8D%7D/uploads/7–30–2012-City_of_Bayfield_Sustainability_Plan_Draft.pdf (accessed March 13, 2013).
22. For information on ecomunicipalities in Wisconsin, see Sustainable Communities Capacity Center. "Wisconsin Communities that Have Adopted Eco-Municipality Resolutions." http://www3.uwsuper.edu/sustainability/Wisconsin%20Eco-municipalities.htm (accessed March 13, 2103).
23. *The New Merriam-Webster Dictionary*. Merriam-Webster, Springfield, MA, 1989.
24. John Friedman. *Retracking America: A Theory of Transactive Planning*. Anchor Press/Doubleday, Garden City, NY, 1973.
25. Ibid.
26. Ralph Keeny. *Value-Focused Thinking: A Path to Creative Decision Making*. Harvard University Press, Cambridge, MA, 1992.
27. Ibid.
28. Carl Jung. http://www.brainyquote.com/quotes/quotes/c/carljung146686.html (accessed March 27, 2013).
29. Zane Maser, personal communication, March 27, 2013.
30. Geoffrey Ball of Geoff Ball & Associates, Palo Alto, CA. Jane had personal correspondence with Geoffrey in 1990.
31. Ibid.
32. Aldo Leopold. "Thinking Like a Mountain." http://www.eco-action.org/dt/thinking.html (accessed March 2, 2013).
33. John Gray. "The Myth of Progress." *Resurgence* 196 (1999):11–13.
34. Ibid.
35. The discussion about Chris's hometown is based on the final articles in this chapter and the public meeting that Chris attended: (1) Aaron Corvin. "Views of Riverbank Plan Offered." *Corvallis Gazette-Times* (Corvallis, OR) September

23, 1999; (2) Aaron Corvin. "Critics Ask for Riprap Halt." *Corvallis Gazette-Times* (Corvallis, OR) September 29, 1999; and (3) Aaron Corvin. "River Plans Face Rough Going." *Corvallis Gazette-Times* (Corvallis, OR) October 5, 1999.
36. Cathy Kessinger. "Scientists Doubt City's Plan." *Corvallis Gazette-Times* (Corvallis, OR) October 6, 1999.
37. The preceding four paragraphs are based on Aaron Corvin. "Consultants Unveil New Plans." *Corvallis Gazette-Times* (Corvallis, OR) October 19, 1999.
38. Aaron Corvin. "Questions Dog River Project." *Corvallis Gazette-Times* (Corvallis, OR) October 7, 1999.
39. (1) An unpublished review of the issues related to the Corvallis riverfront project prepared by a scientist from Oregon State University; and (2) Cathy Kessinger. "Expert: Bank May Be Fine as Is." *Corvallis Gazette-Times* (Corvallis, OR) October 15, 1999.
40. Ibid.
41. Ibid.
42. Ibid.
43. The preceding five paragraphs are based on (1) Aaron Corvin. "Council Plans Another Riverfront Meeting." *Corvallis Gazette-Times* (Corvallis, OR) October 20, 1999; (2) Aaron Corvin. "Consultant Presents Compromise River Plan." *Corvallis Gazette-Times* (Corvallis, OR) October 22, 1999; and (3) Arron Corvin. "Panel Questions City's Handling of River Review." *Corvallis Gazette-Times* (Corvallis, OR) January 20, 2000.
44. Wendy Madar. "Public Must Be Heard Accurately." *Corvallis Gazette-Times* (Corvallis, OR) October 8, 1999.
45. Ibid.
46. Tom Peterson. "Cut Trees So We Can See the River." *Corvallis Gazette-Times* (Corvallis, OR) October 8, 1999.
47. Madar, "Public Must Be Heard."
48. Cathy Kessinger. "Voters Had Faulty Data on Riverbank." *Corvallis Gazette-Times* (Corvallis, OR) October 7, 1999.
49. Unpublished review of the issues related to the Corvallis riverfront project, Oregon State University scientist.
50. The preceding seven paragraphs are based on Madar, "Public Must Be Heard."
51. The preceding three paragraphs are based on (1) Aaron Corvin. "Competitors Kept Out of Bid." *Corvallis Gazette-Times* (Corvallis, OR) October 21, 1999; and (2) "No-Bid Deal Weakens Trust [editorial]." *Corvallis Gazette-Times* (Corvallis, OR) October 27, 1999.

8

Implementing the Comprehensive Plan

> Small actions and choices can have major, although unpredictable effects in determining what comes next. Among the possibilities is that the thousands of experiments and millions of choices to live more consciously will coalesce into a new civilization that fosters community, provides possibilities for meaning, and sustains life for the planet.
>
> **—Sarah Van Gelder, journal editor**[1]

In this chapter on implementing the comprehensive plan, we travel further into the realm of change necessitated by the problems we are facing with current methods of land use control. We look at some of the major issues with zoning ordinances, which are considered a primary tool for the implementation of plans, and then proposed alternatives for the regulation of how land is used. All this is based on the assumption that community planning, done in a way that mirrors nature's inviolable, biophysical principles, will result in less—*not* more—need for regulation. This is not to say that regulations will cease to exist, only that communities can provide incentives, or simply the opportunity for education, to help achieve the same goals as those justifying the existence of land use laws and regulations.

Zoning Ordinances

Zoning law and the policing powers represented by zoning ordinances are a result of state statutes that authorize municipalities to regulate the use of land. Zoning laws are designed to protect the health, safety, and welfare of all citizens.

Zoning is a means for letting people know what they can build and where and what sorts of activities are allowed. Like many laws, there are those who find them intrusive, illogical, threatening, unnecessary, or simply annoying. More often than not, complaints about zoning regulations can be attributed to a lack of understanding of their rational basis. For example, why must there be such restrictions as setbacks or yard requirements, limitations to building height and size, limits on use of property, and so on? While zoning ordinances usually contain glossaries, rarely do they contain the kind of annotation that explains why such requirements as setbacks protect the health, safety, and welfare of the community. Moreover, regulations pertaining to both architectural design and signage are high on the list of complaints about zoning laws, under the general rubric of "arbitrary and capricious."

Public complaint generally ranges along a continuum from annoyance with government regulation in general to anger over perceived violations of property rights. It is no exaggeration to say that violence and near violence are known to occur when property rights are perceived as threatened, particularly in places where property values are high and require excessive debt to own property, frequently with the notion that monetary appreciation is a certainty and thus the debt represents a wise investment. This constellation of factors drives fear associated with perceived rights—fear that education about the broader purpose for development standards may not abate or assuage.

Much of what we hope to engender by the planning process we've proposed, however, is a reduction of misunderstanding leading to fear; strengthening the connection to place and to one another; and the discovery of common ground and values. At the very least, we suggest it is necessary to include an explanatory preface to zoning laws that contains specific explanations and an expanded glossary. Such explanations will help you, the reader, and community citizens to understand the connection of zoning laws to Earth in terms of its biophysical integrity and the sustainability that it engenders, the community in terms of its livability, and the citizens in terms of their health and well-being.

Next, we provide some information that may be useful to include in a zoning ordinance preface or expanded glossary regarding: (1) percentage impervious surfaces; (2) building setbacks; (3) plans for erosion control and contouring the land; (4) open-space requirements; (5) requirements for landscaping; and (6) design controls: site design, architecture, signage.

Percentage Impervious Surfaces

There is a direct relationship between the amount of impervious surface (such as a paved road and the roof of a building) on any one site and its effect on the runoff and infiltration of water. The greater the area covered by an impervious surface, the greater the amount and level of contamination of the water flowing over it from such things as automotive oil in terms of the road and moss killer in terms of a building's roof. In addition, the gutters and storm drains that line paved streets mean that less water is available to infiltrate the soil, where it can be purified, stored, and ultimately replenish the supply of usable groundwater.

There are long-term costs, both social and economic, associated with the pollution of water and an insufficient supply of water. Minimizing the amount of land we cover with impervious surfaces will contribute greatly toward maintaining both the quality and the quantity of our collective water supply.

Building Setbacks

Building setbacks are the distance required for a building or structure to be from a property line. The requirement that keeps property lines free from the encumbrances of buildings is related to access between buildings for the control of potential fires, to protect personal privacy, and to protect the spatial scale of development with respect to such public infrastructure as streets and sidewalks.

Setbacks required from wetlands, lakes, rivers, and streams are in recognition of the impacts human-caused disturbances to the natural landscape have on the quality and quantity of available water. Setbacks from wetlands in particular are in recognition of the social, economic, and environmental values of wetlands with respect to the health of a community; these values include (1) filtering pollutants, sediments, and nutrients from the water passing through them, thereby protecting quality of water in streams, rivers, lakes, and wells; (2) storing runoff from rain and snowmelt, thus helping to prevent flooding and the damage it causes; (3) providing habitat for waterfowl, fish, and other creatures; (4) acting as a shoreline buffer, thus preventing erosion from waves and currents; and (5) providing beautiful open spaces that enhance the quality of life, the economic value of private property, and the economics of tourism.[2]

Plans for Erosion Control and Contouring the Land

One reason for requiring plans to control erosion through the practice of following the contours of the land when moving soil during construction of buildings is to minimize sediment transport by water during and immediately after construction. Sediment transport is a major source of damage to water quality and to the ecological integrity of a community's water catchment in general from myriad sources. In windy Las Vegas, Nevada, on the other hand, airborne soil erosion from disturbed land, both that without and with ongoing human development, contributes to a notable decrease in the city's air quality.[3]

Requiring plans for road grading is also tied to the understanding that the less disturbance there is to the natural landscape, the lower the impact of the development process and finished product on water quality and the health of other natural resources. To the extent possible, conformance with the existing topography is desirable.

Similar to the effects associated with impervious surfaces, there are long-term costs, both social and economic, associated with sediment transport that diminish the quality of a community's available potable water.

Open-Space Requirements

While setback requirements are sometimes spacious enough to qualify as open-space requirements, an open-space requirement frequently stands alone in larger developments. This stand-alone open-space requirement is based on the general understanding, alluded to previously, that the more contiguous and well designed an open-space system is, the better it can be protected in a relatively natural state and the lower will be the impact of a given project on the surrounding ecosystem.

The need for such protection of the landscape is exemplified by the effect on indigenous species between the degree of habitat connectivity and that of habitat fragmentation. Fragmentation of the landscape has resulted in a loss of plant and animal communities, the creation of abundant edge habitat at the expense of interior habitat, the alteration of natural disturbance regimes, and the loss of variability of ecological processes over broad spatial and temporal scales.[4]

The retention of open space can also be justified by understanding the human need for contrast in one's environment, such as quiet spaces to serve as a welcome relief from the built environment. Each type of environment has a sensory impact on humans in that the more intense the development, the more it seems to stress our senses. In addition, provision of open spaces in a cultural built setting typically enhances property value.

Although not in the context of a zoning ordinance, there is increasing evidence of public interest in open-space protection. On Election Day November 2012, voters across the nation gave overwhelming approval for taxing themselves and spending money for new parks and open spaces in their communities.[5] Of the 57 measures on local and statewide ballots, 46 passed, an approval rate of 81 percent. The measures in 21 states will now provide more than $2 billion overall, including $767 million in new money to support protection of water quality, new parks, and natural areas. "From Maine to Texas to San Francisco, we saw voters across the political spectrum say yes to taxes and spending for conservation, which helps their communities," said Will Rogers, president of the Trust for Public Land.[6]

In 2006, the U.S. Forest Service requested public input into the development of the Forest Service Open Space Conservation Strategy and Implementation Plan, focusing on the agency's role in addressing the effects of the loss of open space on private forests; on the national forests and grasslands and surrounding landscape; and on forests in cities, suburbs, and towns.[7] Nearly 9,500 responses were received from the public, and of those received, approximately 80 to 90 percent were overwhelmingly positive and supportive of the Forest Service engaging in the conservation open space.

Requirements for Landscaping

Landscaping requirements are tied to a general understanding that ecosystem integrity, and thus sustainability, can be enhanced or maintained when the landscape surrounding a development is left as undisturbed as possible to protect the ecological processes that not only catch but also allow the infiltration, storage, and purification of storm water. In addition, the larger the percentage of viable populations of indigenous vegetation that remain on a project site, the greater is the cumulative benefit to the integrity of ecosystem processes and services. Thus, landscaping requirements can be seen as a way of protecting or replacing and maintaining something that may be lost in the development process. It can also be seen as a way of beautifying a site, tying it into the surrounding landscape, creating sound buffers, creating a natural cooling system, and so on. Finally, enhancement of a landscape typically heightens aesthetics, human appreciation, and thus property values.

Design Controls: Site Design, Architecture, and Signage

Before offering suggested language for the explanatory preface to the zoning ordinance, with respect to site design, architecture, and signage, we discuss design controls in general—their basis, difficulties encountered with their establishment and implementation, and potential guidelines.

What is design? The dictionary definition that we believe is most relevant to government controls over project design is an underlying scheme that governs how something develops or unfolds and how it functions. We believe a project design must be compatible with, and seem a natural result of, the overall design for a community by supporting the meaning and implementation of its vision, goals, objectives, and policies. We believe further that a design dictates function, much as species composition in a plant community yields the plant community's structure, which in turn yields its function. How well a new project or development functions in a community is tied directly to its design (read: composition and structure). To maintain community integrity, each new development must demonstrate its consistency with the overall design of the community, as revealed in its comprehensive plan. Whereas a good design in this context will be clearly relevant to the overall character of a community by contributing to its social, economic, and environmental well-bring and sense of place, a poor design will seem irrelevant and out of touch with the desires of the citizens.

Design is something that, like planning, is an ongoing process. In planning for a town or our own lives, we analyze trade-offs, consider alternatives, and then choose a course of action. Design emphasizes our physical and spiritual relationship to the world. It can affect how we feel while walking down a street or sitting in a park. How a town looks reveals a community's self-perception. It tells the visitor a lot about who the community is. As Mark Twain once said, "We take stock of a city like we take stock of a man. The clothes or appearance are the externals by which we judge" (p. 2).[8] This is but saying that the definition of good design is important for the development of a community.

Such a definition can be accomplished through published design guidelines that illustrate what requirements are necessary for project approval or for use by peer review committees, such as those described in the following material. Many cities, including Santa Rosa, California, have adopted design guidelines.

Santa Rosa first published design policies in 1980 and updated them in 2010.[9] Although the policies are broad, the city felt that overly restrictive or overly definitive standards would hamper the designer and might result in citywide monotony. Generally, Santa Rosa's site-planning policies cover the appropriateness of setbacks, building height, circulation patterns for traffic, site coverage (percentage of land covered by an impervious surface), parking, and landscaping, which includes outdoor lighting. The policies also cover the relationship of a proposed development to adjacent areas, including developments, the amenities of natural open space, and considerations of shared solar energy.

Its architectural design policies cover scale, mass, bulk, and proportions of acceptable developments; suitability of design to the surrounding neighborhood or the streetscape; consistency of building materials and quality of the workmanship; utility and mechanical facilities and their relationship to other structures; and accessory elements on the site. Sign and graphics design policies address size, scale, location, and materials of such things as signs and billboards with respect to the placement of lighting, proximity to buildings and roads, and so on.

These types of considerations are common in many municipal design standards. A consistent feature in *all* of the municipal design guidelines we have reviewed is protection of the unique character and heritage of place. These municipalities include Santa Rosa; Santa Cruz, California; Capitola, California; Grand Traverse Bay, Michigan; Towamencin Village, Montgomery County, Pennsylvania; Duluth, Minnesota, downtown waterfront; state of Vermont; and Winona County, Minnesota.

There is a human parallel here that cannot be overlooked. Perhaps communities that have discovered the value of protecting their unique character, heritage, and sense of place have discovered the key to sustaining a desired quality of life. Travel writer Arthur Frommer has said: "Among cities with no particular recreational appeal, those that have preserved their past continue to enjoy tourism. Those that haven't, receive almost no tourism at all. Tourism simply doesn't go to a city that has lost its soul" (p. 2).[10]

Many cities, like people cut off from self, have gotten used to mediocrity, disjointedness, humdrum living, or even perceived ugliness; it has become who they are, an identity, so to speak. Many accept this state of being as an inevitable part of progress; for a city, it might be shopping malls, chain restaurants and motels, and characterless, featureless new development; for an individual, it might be a job that provides an income and nothing else. Although the tools to make a city or one's life memorable and beautiful are neither new nor rare, access to them requires deliberate action.[11]

Deliberate action does not always occur, however, until there is some major threat. For a city, this could be the unintended consequences of having a major industry move to a new location—losses not anticipated; for a human, it might be a sudden recognition of one's mortality or some awareness of permanent change in physical strength or flexibility. In any case, a moving force in the direction of self-realization may simply be the full recognition that we do not really miss something until it is gone.

But, how can we help a community understand and identify the images that communicate its values so they can be incorporated into a shared vision? One technique, used by the city of Fort Collins, Colorado, in its comprehensive planning process, is called a Visual Preference Survey. Such a survey was conducted on cable television, in videos, and in public meetings with more than 1,500 participants.[12] While a Visual Preference Survey is considered a useful planning tool for soliciting community attitudes on a range of issues, it does not claim to address all of the relevant issues facing a community during the comprehensive planning process, such as management of growth, transportation, economic development, and so on. But, a Visual Preference Survey is an effective tool for initially involving people in the community planning process on a general and interesting level.

The use of visual images to solicit comments allows people to respond to two questions: Do you like this image? Do you think it is appropriate for Fort Collins (for example)? These questions put the issues in concrete terms that can be readily understood, thereby making it easier than asking the public to spontaneously suggest an acceptable development pattern or density.

Results of this survey may be useful to other communities, even though the sample of respondents was not scientifically drawn, because those surveyed were simply those who choose to be surveyed. The findings, however, may be starting points for discussion in a community wishing to create its own design manual. Participants of this survey in Fort Collins clearly favored a coherent system of open space—including sidewalks, paths, and parks—that connected all parts of the city. Every image that related to a body of water or edges of waterways rated high. What citizens preferred in the built environment were historic buildings—colors, building materials, size, and shape. This, of course, would translate directly to design guidelines for new development in the vicinity of existing historic structures.[13]

Participant preferences for downtown development included mixed-use development with commercial enterprises on the ground floor and professional or residential space above; interesting shop windows; outdoor cafes to stimulate pedestrian traffic; street trees; and street furniture. Preferences for the residential streetscape included narrow streets with on-street parking to slow the speed of traffic, neighborhoods completely connected by sidewalks, and connector streets with planted boulevards to minimize their width and facilitate pedestrian crossing. The images receiving the lowest ratings from a community that values the outdoors included those structures showing an auto-dominated environment, such as streets, parking lots, and shopping areas designed only for the convenience of the automobile.[14]

This kind of an exercise could provide credible substantiation for design standards. Even in the absence of such specific data, it is, in our opinion, still incumbent on a city to offer a rationale for design standards as well as to set forth legal findings necessary for the issuance of a design permit.

Suggested language for use in a zoning ordinance preface might be the following:

> The city of XYZ has established design guidelines for development. The purpose of these standards is to protect and enhance those characteristics that make our community interesting, memorable, and attractive and to protect our unique character and heritage. They are intended to protect all those places, both natural and human-made, that give our community its special character and identity. These include historic buildings, sites, neighborhoods, and natural features of our community that have cultural, natural, or spiritual significance.

There are social, economic, and environmental reasons for design standards as well. If the resulting design enhances people's connection to this place, we anticipate that there will be a greater commitment toward creating a stable economy and stronger forces at work safeguarding our environment. With these things in place, we see a community of strength, resiliency, home, and heart.

Site Design

Good site design has a plethora of rationale to support it. For example, buildings designed and situated to have a strong functional relationship to one another make it easier to collect waste, promote economy of scale in operations in general, and help to form a "legible" site—one that simply makes sense.

To the extent a development can minimize its disturbance to the natural amenities of the site, it is contributing to the well-being of the community. A site's topography and natural amenities must therefore be recognized and used to assist in the design and placement of proposed structures for reasons explained under the discussion of a plan to control soil erosion by fitting a development within the context of a landscape. Another reason is that a community's true economy is realized only when all of its assets are put on the balance sheet. It thus behooves a community to maintain its natural landforms, habitats, edges, waterways, and so on in a viable condition.

Site plans need to be compatible with the immediate environment, including development on adjacent parcels of land and streetscapes. This sense of spatial and temporal compatibility is part of creating continuity and weaving a new development, however one defines it, into the existing landscape.

While an existing landscape might include a strip mall, this does not dictate that new development must necessarily be a continuation of the strip mall design if it is below current design standards. Insertion of new and progressive building design into a dated area is a challenge to the developer and the development-authorizing agency. The opportunity exists, however, for creative solutions that can stimulate rehabilitation and renovation in adjacent developments.

Many derelict neighborhoods in some of the nation's largest cities are being rescued in this manner, usually due to someone recognizing the value, either architectural or social, of reviving a building or a place. Such activity causes a ripple effect, as seen in the renewal of old, derelict neighborhoods in many of the country's large cities. With the improvement of one property, people in adjacent areas see the possibility of creating a critical mass that will give rise to heightened property values overall.

Architectural Standards

Architectural standards, like other design standards, have social, economic, and environmental relevance. If a community has decided, like Santa Rosa, did, that natural building materials, such as wood, stone, or brick, contribute to the character of place, then such a value can be reflected in the community's architectural design standards. To illustrate, the communities of Fort Collins[15] and Santa Rosa saw both social and economic value in having the ground floors of commercial buildings be engaging to pedestrians, such as having interesting display windows, courtyard entrances, provisions for sidewalk cafes or sitting areas, and so on.[16] They also recognized the importance of energy conservation as part of architectural design—a standard that clearly has economic as well as environmental dimensions.

An architectural design standard growing in popularity is that all buildings need to achieve a sense of human scale. This design standard is particularly relevant in places where people recognize the importance of creating nodes of activity or neighborhoods within larger cities, consistent with the approach of New Urbanism.

Signage Standards

Signage standards give rise to some of the most contentious debate in the realm of land use regulation. In some cases, freedom of speech and expression have been added to the sign industry's arsenal in its fight against regulations in general, particularly when it comes to billboards. While states like Vermont and Hawaii have banned freestanding billboards, most communities have either given up and embraced sign proliferation or continue to try to find ways of controlling signs. Increasingly, communities are passing sign ordinances with sunset clauses for the most offensive types of signs, such as billboards or freestanding portable signs.

In an undated publication of the U.S. Small Business Administration in cooperation with the National Electric Sign Association, the question of what your sign can do for your community and neighborhood is answered.[17] According to this publication, signs enhance the environment by making certain zones of the city more attractive and dynamic and giving the area a particular atmosphere. They also perform two major community safety functions: reduction of traffic accidents and crime. Government studies found that vehicle accidents actually decrease at highway intersections sporting commercial signage, and that ambient lighting from signs is a major crime deterrent in urban areas.

The hardest claim for a community to fight that wants to control the proliferation of signs is that they are the most efficient, effective, and consistent revenue-generating devices available to small business. Against such a backdrop of fear of losing one's freedom of expression, claims of public safety, and economic necessity, the sign wars persist.

But, signs, like other design features of the community, must measure up in terms of their contribution to the overall well-being of a community—socially, economically, and environmentally. The placement, number, size, materials, and illumination associated with signs have a collective effect on the overall tapestry of a community and light pollution in the surrounding landscape. Just how prominent a role they should play is up to each community.

Transitioning to Local Living Economy Land Use Practices

Overlay Districts

In the context of zoning law, there is a tool to consider when experimenting with land use practices that align with the biophysical requirements and principles of a local living economy—overlay zoning districts. Overlay zoning is a regulatory tool that creates a special zoning district, placed over an existing base zone, and identifies special provisions in addition to those in the underlying base zone. The overlay district can share common boundaries with the base zone or cut across base zone boundaries. Regulations or incentives are attached to the overlay district to protect a specific resource or guide development within a special area.

A historic protection overlay district, for example, could contain provisions that depart from the underlying district to protect and maintain the historic character of a neighborhood. For example, where the underlying district might call for a 25-foot (7.6-meter) front-yard setback, the overlay district would allow for 10-foot (3-meter) setbacks in recognition of the pattern of development represented by historic structures within the boundaries of the district. This overlay district might also contain architectural design standards for new as well as remodeled or rebuilt structures—requirements that protect the historical nature of the area (e.g., materials, façade design, or color).

Natural resource protection overlay districts can manage development in or near environmentally sensitive areas, such as groundwater recharge areas, to ensure water quality and quantity; to protect and maintain special habitats; or to manage other natural features, such as rock outcrops, riparian areas, or floodplains. Within the natural resource protection overlay, requirements may include building setbacks, density standards, lot sizes, reduction of impervious surface, and specific vegetation requirements. Structural requirements could include such things as building floor height minimums to flood-proof buildings, at least to the high-water level.

Revisions to Development Standards to Accommodate Green Building Practices

The state of Washington provides us with an example for transitioning to full alignment with the principles of a local living economy. Since 2010, a number of cities and counties in Washington State have adopted development standards that support the Living Building Challenge, a building certification process described in Chapter 2.[18] While ordinance language varies in scope and in implementation, all are intended to catalyze advanced, green building practices and define a path through the regulatory process leading to approval of Living Building Challenge projects.

Cascadia, the nonprofit organization sponsoring development of the Living Building Challenge, published two reports in 2009 that were funded by the state and King County; these identified code and regulatory barriers to projects pursuing the Living Building Challenge certification. Washington's Department of Ecology's Beyond Waste Plan (2009) states that a high priority is to continue to "identify and remove regulatory barriers that prohibit and/or contradict green building standards in the State, specifically those related to land use, zoning, stormwater management, water resources and shoreline protection."[19]

In December 2009, the Seattle City Council adopted an ordinance establishing the Living Building Program Pilot Program, which promotes the goals of the Living Building Challenge by providing flexibility in the application of development standards found in the city's Land Use Code. This program limited its application to 12 projects and was renewed for 2 years in January 2013.

According to Jess Harris, Seattle city planner, two pilot projects were approved

> under the Seattle deep green—Bullitt Center and Stone 34 developed by Skanska. [In addition], it's definitely too early to comment on the success or failure of the pilot. However, one could say that it's a success for the movement because Seattle made this opportunity available, promoting

> the [Living Building Challenge] and showing their leadership. Questions about whether the incentive actually works, how well the process works and how we can change code to enable these buildings in the future after the pilot are still outstanding. It's a chicken and egg thing because we really need to test these ideas with actual projects but if we don't get actual projects then it's difficult to determine.[20]

Interested applicants submit a plan demonstrating how the proposed project meets the goals of the Living Building Challenge. Certification is encouraged but not required. The pilot program allows projects that meet at least 60 percent of the challenge's 20 imperatives, plus 75 percent reduction in energy use over a comparable building; 75 percent reduction in water use over a comparable building; and 50 percent of storm water managed on site.

With these as base requirements, an applicant can request departures from land use standards that pose a conflict with standards of the Living Building Challenge. An applicant can request departures from the following standards: permitted, prohibited, or conditional uses; limits to residential density; requirements of the downtown view corridor; floor area ratios; maximum size of use; limits to height; parking requirements; storage of solid waste containers; downtown open space; and so on.

Projects proposed under this pilot program move to the front of the permit line and receive a coordinated staff review. The city established an Innovation Advisory Committee to make recommendations on these projects.

Similarly, Clark County in the state of Washington in 2010 established a pilot program for residential, commercial, and mixed-use projects pursuing the Living Building Challenge certification. Enrollment was limited to a total of six projects, and the program is set to end in 2015. The program was a result of a study aimed at determining code and regulatory barriers for sustainable, affordable, residential development. Bainbridge Island and Ellensburg, two other Washington State communities, likewise adopted legislation to enable the exploration of code revisions to accommodate the Living Building Challenge standards.

Other Regulatory Approaches to Land Use Control

Although there is a strong legal basis for restricting the use of private property, the goal of our proposed revisions to process and outcomes is to have regulations obviated by citizen initiative or the regulations so aligned with community values, vision, and goals that compliance is generally not an issue. With that in mind, there is still a strong case to be made for the protection of legitimate public interests, but we strongly recommend that all legislated land use controls be explained, again in a zoning ordinance preface or expanded glossary, in terms of social, economic, and environmental benefits.

Protection of Farmlands

Conversion of farmland and forests to urban development reduces the amount of lands available for the production of food and timber, as well as the capture and infiltration of water. Soil erosion, salinization, desertification, and other soil degradations associated with intensive agriculture and forestry reduce the quality of land resources and future agricultural productivity. The amount of farmland, forests, and other open space that has been lost to development doubled during the 1990s, according to a government report.[21] Major land use changes have occurred in the United States since the 1980s. The total area of cropland, pastureland, and rangeland decreased by 76 million acres (30,756,109 hectares) in the lower 48 states from 1982 to 2003, while the total area of developed land increased by 36 million acres (14,568,683 hectares) or 48 percent.[22]

A survey of existing government policy, laws, and legal decisions gives credence to the creation of devices for the protection of food-producing lands. The U.S. Farmland Protection Act, for example, states that a "continued decrease the Nation's farmland base will threaten the ability of the United States to produce food and fiber in sufficient quantities to meet domestic needs and the demands for our export materials."[23] The concern represented by this statement, namely, that existing or potential agricultural lands may be undervalued, is real not only because most people living in the United States are so far removed from food production that they have little appreciation of the social value of farmland but also because the outlook for small, private farms seems so bleak with the continued growth of corporate farms.[24]

This being the case, this concern could affect how a community planning committee views its agricultural lands and their future. In addition, local farmers or farming interests, should they exist, could also raise the awareness of community planners.

Boone County, Illinois, for example, protects farmland with a mechanism that kicks in when an application for rezoning from the exclusive agricultural district to rural residential district is submitted. In this instance, no tract of land with 25 percent or greater of its soils classified as class I or II soil or with 50 percent or greater of its soils classified as classes I, II, and III shall be rezoned unless it meets the following requirements: (1) slope of the tract is 6 percent or greater, (2) human-made or physical features act as barriers to farm operations, and (3) it contains at least 20 percent wooded area as judged by the U.S. Forest Service criterion for the evaluation of tree cover.[25]

Such a law has merit in terms of influencing social change because it so clearly relates to a human need for survival—in this case food and the business of farming. Such a law has merit also because a community might not see the need to protect some of its farmlands, believing this necessity can be taken care of elsewhere, such as central California, often termed the "breadbasket of the nation," and thus focus more on growth and revenue.

Regulatory tools for protecting farmland include zoning for a large minimum lot size, fixed area-based zoning, and sliding scale zoning. Large minimum lot size zoning, for example provides an incentive to the property owner for creating a larger lot out of a collection of smaller lots, perhaps 80 plus acres (32 hectares). There could be a tax advantage of owning land with agricultural value instead of the value assigned to land for speculative development. The resources protected through such zoning include agricultural land, rural landscape, and view sheds but would not preclude residential development, only the establishment of residential subdivisions.

Fixed area-based zoning allows the creation of a smaller parcel within a larger one, for example, one dwelling lot of up to 2 acres (0.8 hectares) within a parcel of 25 acres (10 hectares). Like large minimum lot size zoning, the resources protected include agricultural land, rural landscape, and view sheds. With sliding scale zoning, on the other hand, the number of dwellings per area decreases as the size of the parcel increases. Also, smaller parcels not suitable for farming may be developed at a higher overall density than larger farm parcels.

Another example of this is found in a Lincoln County, Oregon, law (Ordinance 1.00075), which addresses rural residential development in agricultural areas. It states that rural residential development can be accommodated on land within agricultural areas not suited for agriculture or commercial timber production. Such residences are allowed if they pose no threat of conflict with prevailing farm and forest practices.[26]

If a community decides that agriculture will not be a part of the community, this value must be respected and portrayed in the plan. As we try to find alternative and more effective means for creating sustainable development, the question that arises repeatedly is whether a community really knows what is best for itself over the long term. If the process through which the community arrives at decisions that affect how the land is developed is inclusive, value laden, and produces multidimensional solutions that serve human necessities while protecting ecological integrity and productivity, the answer, most likely, is yes. The decision-making process is intended to continue beyond the actual plan creation phase and includes mechanisms for monitoring the success or failure of resulting goals, objectives, policies, and laws and their implementation. Thus, if subsequent monitoring reveals that a recommendation is not yielding the desired result, modifications can be made.

Population Growth Rate and New Construction

Unlimited continual growth is not ecologically viable for any community, town, city, state, or nation—hence the necessity of policy and laws. Growth management policies and laws represent the most controversial land use regulation. The city of Carlsbad, California, has determined its future maximum population size by limiting the total number of residential units to 54,600 dwelling units per quadrant and has established both policies and an accounting system to ensure that the limits are observed. To ensure compliance, an accounting system was set up for each quadrant. When individual projects use fewer units than allocated by the General Plan, these units would be held on account in the Excess Dwelling Unit Bank and later could be withdrawn and applied to another project pursuant to established policy.[27]

In Oceanside, California, the municipal code contains a system for controlling residential development. This system, or ordinance, limits the number of building permits allotted. Those permits are allotted on a point system that accounts for such factors as the effect of new construction on public facilities and services, the quality of the architecture and the site for its proposed use, and the desirability of housing for low-income persons or senior citizens.[28] In *Construction Industry Association Sonoma County (California) v. City of Petaluma* (522 F 2d 897, 9th Circuit, 1975), for example, a decision was handed down by the court that builders will receive points for good architectural design, for providing low- and moderate-income dwellings, and for providing various recreational facilities, all of which must conform to general municipal plans and municipal design plans.[29]

Given our reasoning for the necessity of finding innovative ways of controlling and directing development, which includes growth, we believe it may be possible to arrive at a natural cap on development and growth. While pressures from the outside will undoubtedly continue to exist, the laws and policies in place may be so infused within a community's shared vision and the values it embodies that neither the limits to growth nor the availability of resources will be exceeded without further planning or adjustments to accommodate the community's vision and goals. Through such a process, innovative solutions can be forged to either stem growth or reconcile more growth with a community's immediate landscape and its environmental constraints, both of which depend on what a particular community desires.

The following examples of attempts to regulate growth are taken from my own (Jane's) experience in two cities where growth management was a major focus of planning efforts. In Santa Barbara, California, in the 1970s, growth became a major election issue and swept in a new majority to the city council. It was an "antigrowth" majority who directed the planning department to devise techniques for regulating growth following a citizen referendum that called for a population cap of 75,000. One technique that we proposed was to change the regulation of population density from units per acre (hectare) to bedrooms per acre (hectare), with the assumption that bedroom occupancy was generally 1.3 persons per bedroom in the typical household. This would give developers more flexibility with number of units while giving the city greater assurances of not exceeding the population cap.

In Santa Cruz, where a county-imposed limit on the annual number of building permits issued forced most of the development into cities, primarily Santa Cruz, an attempt to regulate growth took the form of regulating the size of a household. The proposal required owners of rental dwellings to obtain a commercial-residential permit. The permit was issued if the property owner agreed to a number of conditions, including (1) allowing only a specified number of residents to occupy a dwelling in relation to the number of defined bedrooms in the dwelling and (2) allowing only a specified number of off-street parking spaces. Since more than 50 percent of the dwellings in Santa Cruz were rentals, this was essentially a form of population control.

In another attempt to have greater control over population growth while attempting to increase housing affordability, a city of Santa Cruz housing task force proposed changing the multifamily residential density standard from units per acre (hectare) to population per acre (hectare). The intent was to encourage the development of a greater number of smaller units but it also clearly was another means for managing growth.

Such mechanisms are usually weak when it comes to actually preventing growth, as revealed in both Santa Barbara and Santa Cruz, which continue to experience tremendous pressure for further development primarily because they are both beautiful towns with highly desirable climates.

Nonregulatory Methods of Controlling Land Use

There is an array of nonregulatory tools available for managing our use of the land, such as processes initiated by citizens groups and specific laws that enable and provide incentives for citizens to initiate actions aimed at protecting the land. We begin our discussion with incentive-based tools.

Incentive-Based Tools

Incentive-based tools, which are described in great detail in a number of publications, such as *Common Groundwork: A Practical Guide to Protection Rural and Urban Land: A Handbook for Making Land-Use Decisions,*[30] include (1) outright purchase of land, (2) donated conservation easements, (3) purchase of development rights, and (4) transfer of development rights.

Outright Purchase of Land

The outright purchase of land from a willing landowner by a government agency or a not-for-profit conservation organization is generally for the purpose of protecting natural resources or providing public access. An example might be private land within a national forest or national park. On the other hand, a landowner may choose to donate some amount of land to a government agency or a not-for-profit conservation organization and reap tax advantages in return.

Donated Conservation Easements

Donating conservation easements enables landowners to permanently limit future development on their property. Such an easement is a voluntary legal agreement between the landowner and a public agency or a not-for-profit land trust. Under such an arrangement, the landowner retains both the ownership of and the right to use the land, but within the terms of the conservation easement. There are, of course, tax advantages associated with such easements.

Purchase of Development Rights

Purchase of development rights is a government-initiated or not-for-profit-initiated program that pays a landowner (who wants to protect his or her property from the potential of future development) for the development rights. Purchase of development rights takes the form of a legal agreement that places a permanent conservation easement on the properties.

Transfer of Development Rights

Transfer of development rights, on the other hand, is a program that creates a private market for buying and selling the right to develop property. A transfer of development rights program begins by dividing a community into "sending zones" and "receiving zones." Sending zones are areas that a community wants to protect from further development. Receiving zones are areas that are able to accommodate new growth, usually because they are already served by existing infrastructure, such as sewers, roads, and water lines. In such a process, the right to develop the land is separated from other rights associated with property ownership, such as privacy and the right to lease or sell the land. Property owners in a sending area sell development rights to developers to use in receiving areas, where higher-than-normal densities are typically allowed.

Such programs are complicated to administer but provide advantages to the community at large as well as to program participants. Advantages include the guidance of development to areas that can most easily support it and the protection of lands that cannot. "In addition, local governments do not have to spend funds to acquire land or development rights. In contrast to cluster zoning, which protects land parcel by parcel, Transfer of Development Rights can help to preserve large tracts of open space in the community."[31]

The number of possible incentive-based tools is probably limited only by the resourcefulness and creativity of individuals and groups interested in furthering responsible land use. It is through an understanding of community values, the multidimensional requirements of both humans and the ecosystems on which they depend for personal and social survival, that releases the creative forces of experimentation and discovery with respect to incentives for wise use of the land.

Many cities offer *density bonuses* in trade for the developer providing affordable housing, parks, improvements in infrastructure, child care facilities, public art, and so on. In Santa Barbara, a density bonus is available related to the size of the dwelling. That is, decreasing the size/square feet (meter) of a unit can allow for increased density.

Incentives Themselves

Most incentives are based on financial savings or even gain and have proven useful in achieving better land use practices. Tax advantages, actual payment to private landowners (in the case of purchased development rights), and allowing higher building densities, which makes a development more profitable, are examples of such incentives.

Ideally, incentives would not be needed to achieve responsible land use practices. Land trusteeship could result from education, inclusive government process, and recognition or rewards for leadership in this area. In the late 1990s, the city of Ashland, Wisconsin, held a community event to celebrate years of citizen-driven work to revive the life of its waterfront and make it more accessible. The purpose of this celebration was threefold: (1) to recognize and acknowledge that the community has something to offer the city in developing its waterfront; (2) to recognize and acknowledge the current success in redeveloping the waterfront as a form of reward that, it is hoped, will instill friendship between the government and the citizens and stimulate further contributions to civic and community life; and (3) to draw, from the community, its memories, stories, and photographs relating to the waterfront, which has had a rich and colorful history.

This information was used to design a series of interpretive signs to complement the recently completed waterfront trail, which travels from one boundary of the city to the other—a distance of about 5 miles (8 kilometers). The signs contain information about historic sites; existing industry along the waterfront, such as the power plant and international paper-folding equipment plant; the types of fish caught in the lake; the history of the largest remaining ore dock on the Great Lakes; special seabird nesting areas; and so on. Inviting citizens to share their stories, memories, and photographs is a form of recognition and reward in and of itself for simply being longtime residents in a particular place, which is exactly the kind of activity that strengthens community.

Another example of this occurred during the 1990s in the town of Washburn, Wisconsin, when the town decided to apply to the Minnesota Design Team to conduct a 3-day charette to help the community discover ways of improving its town; in this case, the charette drew around 200 citizens or about one-fourth of the community. The 17-member Minnesota Design Team was composed of volunteers—planners, architects and other design professionals, engineers, sociologists, and landscape architects—who, just for the cost of travel expenses, visited communities that desired their assistance. Local families hosted team members on location.

Before their arrival, the Minnesota Design Team sent a manual to the community of Washburn that explained what sort of community seeks the team's assistance. It is a community committed to the future, a community that wants to make the most of the biophysical components of its environment as they define the community. But, it is also a community that realizes no one person or group of people are in and of themselves wise enough to know the best course of action for the community's future. The community therefore sets out to involve as many of its citizens as possible in the creation of a shared vision of the future that incorporates the ideas and core values of the community as a whole. To crystallize ideas, to clarify core values, and to energize the community effort, an outside group of eyes, ears, and minds is invited to help them—the Minnesota Design Team.[32]

The experience for most people involved with the Minnesota Design Team was positive since the team was well trained in active listening. The value they placed on the contributions of all participating citizens was a form of reward that has had lasting value to the community of Washburn in the form of heightened citizen participation in the affairs of government and citizen-initiated projects, such as protection and enhancements to the city's waterfront trail, which brings us to the notion of a development review.

Development Review

There are many opportunities for producing the type of development that does not compromise an ecosystem's ability to deliver services that sustain life. The most proactive level is during the community planning process, when the vision and goals are created, tied to a community's core values, and interwoven into a review of the potential effects on the ecosystem caused by proposed actions. This process, of course, should yield zoning laws and land use policy consistent with the community's vision and goals. And, this process may indeed stimulate citizen initiative, at both the individual and organization levels, aimed at reaching these goals.

Citizen-Initiated Development Review

The Conservation Fund proposed a point system for development. This system was in response to what was perceived as mounting pressure on developers to produce more environmentally friendly developments.[33] The Conservation Development Evaluation System is a rating system created to evaluate the effects of the development on such things as long-term water quality and landscape integrity over the development's lifetime.

The Conservation Development Evaluation System's main purpose is to prompt developers to think about environmental concerns earlier in the planning process. The system is intended for use in rating new and existing developments and is to encourage developers to strive to meet rigorous standards and achieve multiple ratings for their developments. It therefore neither is a piece of legislation nor purports to set the stage for recommended legislation. Whereas this constitutes the strength of the Conservation Development Evaluation System, its weakness is in the same area as that of the Land Development Guidance System of Fort Collins: its emphasis on points, which are awarded or deducted for satisfying a specified criterion.

Some of the ratings of the Conservation Development Evaluation System are much more stringent than the norm. The ratings of the Conservation Development Evaluation System lie more in the realm of positive feedback loops, however, and thus are more likely to change the behavior of people than is the legislated, but optional, point system of Fort Collins.

In contrast to the system in Fort Collins, however, this system presents a method of evaluating "conservation developments"; recognizes the site-specific constraints and limitations on planning and design; assumes that the predevelopment condition of the site is usually disturbed; is not intended for undisturbed land or high-quality natural areas; acknowledges that each category may not apply to every development; and emphasizes impact on water quality. The core criteria by which each development is rated are (1) site design and construction practices (percentage impervious surfaces relative to conventional development, preservation of natural features or change in land form, sediment and erosion control); (2) storm water management (rate and volume of runoff); (3) open space (management of open space, environmentally constrained open space); and (4) protection of natural resources (development of protection plan for natural resources and existing vegetation, such as native plant and tree conservation or newly planted vegetation).[34]

Certified forestry products also represent the tactic of awards or incentives for encouraging sustainable development. These are wood products produced by a system that has been certified for sustainable practices, from logging and milling to manufacture and distribution. The market for certified wood products is growing as consumers become more familiar with the practice of sustainable forestry.[35] As these markets and their attendant demand grow, it is likely that more and more businesses associated with forest products will pursue these markets and thus change their behavior toward the environment. Be that as it may, for some consumers there is a certain status associated with owning what might be viewed as a designer wood product—SmartWood products do carry a label, much like many designer clothes.

Endorsements, like certification, represent a positive feedback loop in a system. The development of such a process took place in 1999 in Traverse City, Michigan.[36] The peer review committee of the Traverse City Area

Chamber of Commerce New Designs for Growth program was comprised of architects, civil engineers, realty agents, surveyors, developers, home builders, and conservationists. All shared an interest in seeing that principles of good growth management and progressive community development were applied to any project with potential significant impact.

After 2.5 years of existence, the review committee became so influential with community planners, government officials, and developers that its endorsement often ensured friendly reception of project plans by planners. By 1999, all but 2 of its 25 projects reviewed received final approval. The chamber also produced a design guidebook for use by developers. The mission represented by the guidebook was "to preserve and enhance the quality of our natural resources and environment as the basis for a healthy economy."[37]

A statement by Keith Charters, coordinator of New Designs for Growth and administrator for the review committee, demonstrates an understanding of the use of leverage points in a system to achieve change. "Bad development is not the fault of developers, contractors or realtors. It is the fault of bad rules, rules that affect the layout, look, site usage and signage of projects and that specify where they can go." In addition to the peer review committee, New Designs undertook a massive education program, holding workshops for officials and residents in 47 of the 93 jurisdictions of the region. This committee was also seen as a stopgap mechanism to help shape and control growth until local communities update their master plans and zoning ordinances.[38]

The work of the Traverse City Area Chamber of Commerce New Designs for Growth peer review committee is laudable on a number of counts: (1) It was citizen based; (2) it was nonregulatory; (3) it had a strong educational component; and (4) it attempted to intervene at nearly the highest level for success—to change goals in the community. We also believe that such an organized effort emerges from a process that is inclusive, vision driven, tied to clearly defined goals, and high profile.

Such an organized effort brings the citizenry to government on a regular basis and, if carried out in an emotionally safe, civil, and dignified manner, can provide a productive adjunct to the government process. Planning commissions can work well if they carry out a citizen-based independent review process, which means the appointed members do not feel obligated to carry out the political agenda of the person who appointed them—be it the mayor or a city council member. This is not to say that citizens on a peer review committee will not have personal agendas, but rather that a citizens' committee can choose to operate more freely—without the legal and political constructs of a government-appointed body.

Such a multi-interest citizens' group has both advantages and disadvantages, as does everything, but on balance represents the essence of democracy and democratic processes and therefore represents an opportunity to be seized.[39] Advantages of such a group includes its diversity of interests—a characteristic of stable systems that creates an arena in which participants can simultaneously address the interconnected aspects of a project. In addition, the educational value embodied in the process will increase the participants' knowledge of the issues discussed.

Disadvantages of such a group include the fact that their work, due to the consensus process, often takes more time than in processes in which decisions are based on a vote of the majority. Nevertheless, the consensus process generally leads to more durable decisions.

Obstacles to Development Review

Revised or new land use policy could take the form of a new or revamped development review process. This process, if consistent with the type of planning we are advocating, would have a strong component of community involvement. Public participation is, after all, as important at the planning level as it is at the stage at which significant development is being proposed.

Some governments look at community involvement as a threat or obstacle to new development since many developers do not want to invest the additional time needed for development approval. This is a business decision since time is money, and time spent for adequate community review or government review that does not ultimately allow development is seen as a poor investment of time, which equals a loss of money.

Allowing public participation in the review process recognizes the importance of choice and revering its place in civil society. Choice—a simple but powerful word—changes the world every day in ways both great and small. We value it, vie for it, and die for it. Yet, many people have little or no concept of the sense of freedom that comes with the ability to choose or the sense of responsibility that comes with the consequences of one's choice. Thurgood Marshall and Mahatma Gandhi understood it. They made no apologies for wanting it equally for themselves and for all other people. They did much to make the world see that individual choice—implied by basic human rights—has no meaning unless it is universally available to all people. The ability to choose confers on each individual a sense of value, self-confidence, and the dignity of being human.

In local living economies, citizens guide the sustainable development of their own community within the mutually sustainable context of their landscape. After all, the local people and their children must reap the consequences of any decisions that are made. To limit their choices is to force someone else's consequences on them, often at a great and increasingly negative long-term cost, first socially and often environmentally.

In 1999, builders and developers in Oregon devised a way to directly intimidate people opposed to their commercial developments, effectively rendering silent any opposition at public meetings held to discuss their planned enterprise. Such cases are known as strategic lawsuits against public participation or SLAAP suits.[40]

The Oregon House Judiciary Committee reviewed House Bill 2805 (aimed at stopping SLAAP suits) in March 1999 after receiving statewide testimony concerning the abusive lawsuits. During the testimony, witnesses said they had ended up having to defend themselves in court after voicing their opposition to various proposals dealing with the use and development of land.

Although House Bill 2805 would protect citizens from lawsuits as a result of presenting facts, stating opinions, or expressing feelings at public meetings, it would not protect them from such statements made in letters to newspapers, such as letters to the editor. The bill also provides for courts to award attorneys' fees, other costs, and even punitive damages to defendants winning a suit under the bill's provision and provides immunity for testimony given in public meetings.

Backers of the bill were pleased on May 12, 1999, when the Oregon House passed it by a wide margin because, as Evan Manvel from 1000 Friends of Oregon, said "Public participation is critical to statewide planning and Oregon democracy." Liz Frankel, from the Corvallis League of Women Voters, agreed but cautioned that, while Joe Chandler (a lobbyist for the Oregon Building Industries Association, a statewide home builders organization) did not push very hard in the House, he would in the Oregon Senate.[41]

Although the Oregon House passed House Bill 2805 in May 1999, it was severely altered and then defeated in the Senate in July of the same year. Jeff Lamb, chairman of Oregon Citizens for a Voice in Annexations, said the defeat of the bill "will just add to the cynicism and the apathy" that already exist among voters. "It will be ... a very chilling denial of what Oregon prides itself on ... citizen involvement."[42]

Representative Kurt Schrader, a Democrat and the bill's sponsor, criticized the Senate action, which he contended put an end to the "opportunity for lawmakers to reaffirm the right of citizens to directly participate in the democratic process without fear of retaliation." "It's awful hard to go against money," Lamb said, which, according to Lamb, proved that, "He who has the campaign contributions will make the rules and the little guy will not have any immunity for participating in public debates."[43]

The Senate vote on the bill was along party lines to table the bill, with the Republicans engineering the defeat. Republican Senator Neil Bryant defended reducing protections for people testifying at public hearings on grounds that speakers need to do the research necessary to make sure their testimony is accurate, which deftly removes social values and the feelings they engender from the democratic process.

Although all cities in Oregon must create citizen groups to involve people in land use planning, according to state land use laws, in Corvallis the planning commission filled that role prior to the formation of the Corvallis Committee for Citizen Involvement, which was intended to act as a neutral clearinghouse for information. While a city can legally choose its own planning commission to fulfill the requirements of the state's land use law, it is not recommended.

The Corvallis Committee for Citizen Involvement seeks answers and invites residents to help search for ways of getting citizens involved early in the development process—before ground is broken and houses go up. An option to consider might be the addition of an ordinance that requires developers to meet with neighbors *before* submitting development plans.

The Corvallis Committee for Citizen Involvement also developed a series of brochures that illustrated how land use decisions are made and how citizens can give effective testimony during public hearings. The committee also studied how it might propose alternative public notices of land use proposals that were more reader friendly.

Fiscal Impact Analysis: Require for Specified Development Proposals

Decision makers concerned about thwarting the development process and potential growth in the community due to a development review process that is open to community debate would be best advised to consider not only the collective intelligence of the citizens but also the right of citizens to participate in creating their own future. At the same time, however, a city would most likely benefit from conducting a fiscal impact analysis on all new proposals.

A fiscal impact analysis can provide a picture of the costs and benefits of a land use project, and the following questions can be answered:

1. Does the proposed business have a business plan?
2. Are options for diversification built into the plan?
3. How many employees will be needed, both short and long term?
4. What are the hiring practices of the business?
5. How many jobs will this business mean to the local community?
6. What sorts of adjustments will be needed in city infrastructure to accommodate this amount of growth?
7. What will be the impacts, if any, on police, fire, and waste collection services; on the capacity of the sewage and water treatment plant; and so on?
8. Will there be an impact on local schools and the available supply of housing?
9. Will there be a short-term gain for the community with a long-term loss, or will it ultimately be awash in terms of revenue added to the city?

A modified fiscal impact analysis could incorporate an environmental impact analysis, which analyzes the interconnectedness of three kinds of capital: social, economic, and natural. Although a traditional fiscal impact analysis, as described in the series of questions, does assess some social factors, such as schools, jobs, and housing, it only marginally touches on environmental effects by assessing the impact of a development on waste collection services.

If we assess the true cost of bringing a new development to a community, we would consider the long-term social, environmental, and economic costs, as well as the monetary gains. For example, dealing with potentially polluting emissions spewed into the air or water can be costly, both economically and socially, and thus has merit for the company as well as the community.

In evaluating the overall impact of new development, some judgment calls will certainly be required simply because we cannot know what all the effects might be, including a potential array of unintended consequences. The best we can do is to have in place an evaluative process that gives us a credible measure of potential consequences with some degree of certainty before approving a new development. Further, in judging whether a new development is "right" for one's community, there must be indications of the overall integrity of the business. This can be assessed from the existing track record of the business in question, from the business plan and whether it contains measures for diversification, and most important, whether the business is willing to accommodate the community process aimed at measuring how well the business fits into the overall vision and goals of the community.

While working as a consultant to a city in northern Wisconsin, I (Jane) witnessed an interesting interaction between a company wanting to come to the community and the community's response. The company approached the city government about locating its business in the city's industrial park. This business, which promised 60 local jobs after 5 years of operation, was the production of a building material created through the incineration of garbage. This high-intensity incineration process burned the organic material in garbage and left a residue cast into pellets, which were used in the manufacture of lightweight, durable concrete building blocks. To make the operation profitable, truckloads of garbage would have to be shipped in daily to add to the waste collected from the city.

Much to the credit of the city, it held a public information session so the company could present its case to the community. The session, although not required by law, was seen as giving the citizens an opportunity to comment on a project that had already raised serious concerns because of the potential environmental consequences. Although there would have been a required state and federal environmental review process for such an operation along the shore of Lake Superior, this early review enabled community values to be aired, which ultimately led to the company going elsewhere.

If a traditional fiscal impact analysis had been done on this project, it is hard to say what the bottom line would have looked like. But, from a sustainability perspective, this business would have rated poorly because of the necessity of importing garbage from outside the area with the attendant heavy truck traffic in addition to the air pollution from the high-intensity incineration process.

As a result of this project having reached the public arena, a neighboring tribe of indigenous Americans decided to apply for a federal class A air status, which, if received, would preempt any neighboring jurisdiction from allowing development that would require a special permit to operate because of the air pollution it would cause. This action in turn created a rift between the city and the tribe, which remained unresolved and unmitigated for years.

The point of the story in the context of development review is not only to demonstrate the importance of a public review early in the process but also to ensure that the input is received from a representative cross section of the community. In this case, because of the regional effects of the manufacturing process, the local community, as well as leaders from adjacent communities, should have been invited to offer their comments, feelings, and values, which brings us back to a question asked previously: Does a community know what is best for its future?

Did the community really know that this industry would be damaging to the environment since a thorough analysis had not been obtained? Would not the burning of garbage to create another product be consistent with the ecological principle whereby all waste is fuel for other living systems? If the city had had in place a process for evaluating potential new development in the community that itself was a product of community-wide planning, a number of things could have been avoided, such as the rift between two government entities, a certain wariness in the community about the kinds of industry the city might be considering, and lingering concerns about the lack of predictability for citizens and developers alike with respect to how potential new development is evaluated.

Environmental Impact Analysis

Currently, state laws in most states as well as federal law require environmental impact assessments if a proposed project is a public utility or if there is potential for negative consequences to the quality of the air or water. In some states, like Florida and California, state laws require environmental review for almost all commercial and industrial developments, as well as residential projects that exceed a certain size.

The review process in California starts with an Environmental Checklist, which is part of an initial study.[44] The checklist poses a series of questions relating to the potential impacts on soils; air; water; plant life; and animal life; as well as those relating to noise pollution; light pollution; land use; natural resources; risk of upset (e.g., risk of explosion or the release of hazardous substances in the event of an accident); population (location, distribution, density and/or growth rate of the area's population); housing; transportation/circulation; public services; energy; utilities and service systems; human health; aesthetics; recreation; and cultural resources. It also calls for compliance with the community's general plan—a good place to check for consistency with the community's vision. Although local municipalities are free to devise their own format for initial studies, the completion of the Environmental Checklist represents a specific determination that must be made regarding whether a proposed project may create enough potential negative consequences to require further analysis.

While this tool has much merit, it lacks dimensions of the community that could potentially be affected by major new development, which leads us to the notion of a checklist for social-environmental sustainability.

Checklists for Sustainability

Checklists for sustainability, also discussed in Chapter 3, have emerged mainly from nongovernment organizations during the last decade because they are, by their nature, nonquantitative and therefore allow too much room for legal contest. If a decision-making body, whether governmental or nongovernmental, is clear on a community's vision and goals, however, the use of such a checklist is much easier than if the vision and goals are not in place.

While this rather complex review process would discourage some new interests from coming to a community, we maintain that if a community is desirable enough, this sort of hurdle would not discourage developers with similar standards who would therefore understand the long-term benefits of such standards to the community as a whole. Let us look at some examples of these lists.

The Sustainable Community Checklist, generated by the Northwest Policy Center at the Graduate School of Public Affairs, University of Washington, is designed to help communities assess their own level of sustainability.[45] This workbook presents six principles of sustainability to help communities think more holistically about the relationships among economy, community, and environment. Each of the six principles is accompanied by a corresponding set of indicators and examples of projects in the state of Washington that relate to the principle and indicators.

The workbook offers a caveat that no precise blueprint exists for what sustainable communities will look like. The vision, and the strategies needed to get there, will vary from place to place. Like the checklist described previously, this checklist suggests a way of evaluating how well a proposed new development would fit into a community guided by principles of sustainability. The principles offered by this workbook point out that sustainable communities foster commitment to place, promote vitality, build resilience, act as stewards, forge connections, and promote equity:

1. In terms of fostering commitment to place, the first principle poses the questions of whether a business offers living wage jobs and has demonstrated, or at least voiced, its understanding of the community and its vision and goals.
2. The second principle, which addresses community vitality, considers the extent to which a community stresses conscious development

over continual growth and encourages local businesses to provide products that are not currently available to keep dollars circulating within the community.

3. The third principle, resilience, might be reflected in a new business by the extent to which it contributes to the diversity of businesses in a community, whether or not diversification is built into its business plan, and its general ability to sustain itself.
4. The fourth principle, which addresses stewardship, posits whether a proposed new business expresses an understanding of efficient use of resources and responsible waste management.
5. The fifth principle, relating to the creation of connections, might be revealed in a new business by the extent to which it proposes to relate to existing businesses for mutual benefit.
6. The sixth principle, promotion of equity or fairness, could be demonstrated by the company's hiring practices or its view of the future and how its product, services, or operations consider future generations.[46]

A community can use checklists of the types discussed in a number of ways. The foregoing discussion applied them primarily to new business and their use in evaluating the fit of a new business with one's community. These checklists can be adapted, however, for new development in general.

The Traverse City Area Chamber of Commerce New Designs for Growth peer review committee, described previously, might use such a checklist. Inserting checklists such as these into the government decision-making process would mean revising zoning ordinances in a way that would (1) base the approval of certain types of development on how well a proposal ranks according to such a checklist, (2) develop findings for permit approval that are tied to components of such a checklist, and (3) simply combine such a checklist with a standard environmental impact checklist and include it as part of the preliminary review of a project. Further, provisions under a zoning ordinance would have to be clear on what sorts of projects merited this level of review.

Zoning ordinances set forth permitted uses that require only administrative approval, which can typically be carried out by a zoning administrator or planner. Other uses usually fall into a category of either a special use permit or a conditional use permit. It is at this level that the public is typically permitted, officially, to participate in a development review through the public hearing process. It is at this level of review that a checklist, such as we are describing, would naturally fall. On the other hand, there could be a special category of *uses* that always require this additional review.

While municipalities have broad latitude in prescribing conditions for approval of development, careful legal scrutiny is important, of course, to avoid litigation. Some communities have become such desirable places to live and work, however, that even the most stringent review does not seem to diminish the demand for development. Santa Barbara is one such place. This town seems to be caught in a self-reinforcing feedback loop that the more this place works for the people living and working there, the more it "works" and, as such, is nearly being drowned by its own success, which led to an intensive study in Santa Barbara to discover leverage points in the system to control the city's growth and its negative effects, which brings us to monitoring in the next chapter.

Summary

Chapter 8 examines the best means of implementing a comprehensive land use plan. Chapter 9 is a comprehensive look at how to monitor the progress of a land use plan once it is implemented.

Endnotes

1. Sarah Van Gelder. "Fourth Set of Indicators." Sustainable Seattle. n.d. http://www.sustainableseattle.org/programs/regional-indicators/124 (accessed March 5, 2013).
2. Many of the relationships in this section were drawn from (1) The Conservation Fund. *Conservation Development Evaluation System/Revised Draft*. Arlington, VA: Conservation Fund, 1999, 10 pp.; (2) *Natural Resources Code 117*. Madison, WI: State of Wisconsin, 1999.
3. Personal communication between Chris and Alan Pinkerton, a planner with the Clark County Department of Comprehensive Planning, Las Vegas, Nevada.
4. Robert G. Lee, Richard Flamm, Monica G. Turner, and others. "Integrating Sustainable Development and Environmental Vitality: A Landscape Ecology Approach." In: *Watershed Management*, Robert J. Naiman (ed.), 499–521. New York: Springer-Verlag, 1992.
5. "Voters Give Landslide Approval for Conservation Spending Nationwide." November 7, 2012. http://www.tpl.org/news/press-releases/2012-press-releases/voters-give-landslide-approval.html (accessed March 3, 2013).
6. Ibid.
7. U.S. Forest Service. "Executive Summary of Public Input." n.d. http://www.fs.fed.us/openspace/Executive%20Summary%20of%20Comments%20from%20the%20Public_final_website.pdf (accessed March 2, 2013).
8. Edward T. McMahon. "Community Appearance and Tourism: What's the Link?" *Heritage Tourism Update* Winter (1993):2–3.13.
9. City of Santa Rosa, California. "Design Review Policies." n.d. http://ci.santa-rosa.ca.us/departments/communitydev/development/Pages/DesignReview.aspx (accessed June 4, 2013).

10. McMahon, "Community Appearance and Tourism."
11. Ibid.
12. City of Fort Collins. "Results of the Visual Preference Survey." n.d. http://www.fcgov.com/advanceplanning/pdf/vps1-doc.pdf (accessed March 6, 2013).
13. Ibid.
14. Ibid.
15. Ibid.
16. City of Santa Rosa, "Design Review Policies."
17. Small Business Administration in cooperation with the National Electric Sign Association. "Signs and the Small Business—Focus on Facts." n.d. http://www.signdesigninc.com/uploads/pages/FocusonFacts.pdf (accessed March 6, 2013).
18. This discussion based on King County Green Tools. "White Paper: Local Ordinances Related to the Living Building Challenge." September 2012. http://living-future.org/node/351 (accessed March 3, 2013).
19. Department of Ecology, State of Washington. "Beyond Waste." n.d. http://www.ecy.wa.gov/beyondwaste/ (accessed March 6, 2012).
20. Jane's personal correspondence via e-mail with Seattle city planner Jess Harris on March 5, 2013.
21. Philip Brasher. "Development of Farms, Open Space Doubles during 1990s." *Corvallis Gazette-Times* (Corvallis, OR) December 18, 1999.
22. JunJie Wu. "Land Use Changes: Economic, Social, and Environmental Impacts." *Choices: The Magazine of Food, Farm, and Resources Issues* 23 (2008):6–10. http://www.choicesmagazine.org/magazine/pdf/article_49.pdf (accessed March 5, 2013).
23. American Farmland Trust. "Farmland Protection Policy Act of 1981." http://www.farmland.org/about/mission/documents/AFT_FPPA_6–04.pdf (accessed March 7, 2013).
24. Steve Apfelbaum and Jack Broughton. "Applying an Ecological Systems Approach to Urban Landscapes." *Land and Water* January/February 1998. http://www.appliedeco.com/Projects/ApplyiEcolSystems.pdf (accessed March 5, 2013).
25. To gain a sense of how Boone County protects farmland, see "Ordinance Adopting and Implementing the Boone County Agricultural Conservation Easement and Farmland Protection Program." Ordinance No. 06-18. 2006. http://www.farmlandinfo.org/documents/30944/Boone_Co._Ill.__Ordinance_06-18.pdf (accessed March 7, 2013).
26. Lincoln County, Oregon. "Lincoln County Code, Chapter 1: Land Use Planning." http://www.co.lincoln.or.us/counsel/code/2012_code/2012-LCC-Chapter-01.doc.pdf (accessed March 7, 2013).
27. City of Carlsbad, California. "Growth Management." n.d. http://www.carlsbadca.gov/services/departments/planning/pages/growth-management.aspx (accessed March 5, 2013).
28. City of Oceanside, California. "Current Codes and Regulations." 2013. http://www.ci.oceanside.ca.us/civica/filebank/blobdload.asp?BlobID=25428 (accessed March 7, 2013).
29. *Construction Industry Association Sonoma County v. City of Petaluma*. August 13, 1975. http://aalto.arch.ksu.edu/jwkplan/Briefs4/petaluma.htm (accessed March 7, 2013).

30. Joseph H. Chadbourne and Mary Chadbourne. *Common Groundwork: A Practical Guide to Protecting Rural and Urban Land: A Handbook for Making Land-Use Decisions*. Institute for Environmental Education. American Planning Association Planners Book Service, Chicago, 1993.
31. Ibid.
32. Minnesota Department of Trade and Economic Development. *Governor's Design Team Community Manual*. St. Paul, MN: State of Minnesota, 1991.
33. Personal communication between Chris and Alan Pinkerton, a planner with the Clark County Department of Comprehensive Planning, Las Vegas, Nevada.
34. Many of the relationships in this section were drawn from (1) Conservation Fund, *Conservation Development Evaluation System*; and (2) State of Wisconsin, *Natural Resources Code 117*.
35. Robert Brander, SmartWood Coordinator, Upper Midwest. Telephone conversation with Jane Silberstein, 1999.
36. Bill Echlin. "Peer Site Review Committee Responsible for Guiding Growth." *Record-Eagle* (Traverse City, MI) January 10, 1999.
37. New Designs for Growth. "GuideBook." 2008. http://www.newdesignsforgrowth.com/pages/guidebook/ (accessed March 8, 2013).
38. Bill Echlin, "Peer Site Review Committee."
39. The following discussion of multistakeholder groups is based on Environmental Defense Fund. *Environmental Sustainability Kit*. New York, NY: EDF, 1996.
40. The following discussion of SLAAP suits is based on (1) John Butterworth. "House Passes Legislation to Stop SLAAP Suits." *Corvallis Gazette-Times* (Corvallis, OR) May 13, 1999; and (2) "Anti-SLAAP Legislation Defeated." *Corvallis Gazette-Times* (Corvallis, OR) July 7, 1999.
41. Ibid.
42. Ibid.
43. Ibid.
44. Arron Corvin. "Committee Seeks to Encourage Public Involvement in Local Land Development Process." *Corvallis Gazette-Times* (Corvallis, OR) December 8, 1999.
45. Rebecca Bauen, Bryan Baker, and Kirk Johnson. *Sustainable Community Checklist*. Northwest Policy Center, Graduate School of Public Affairs, University of Washington, Seattle, 1996.
46. Ibid.

9

Monitoring Progress

> We believe that the sustainable community will be promoted where many players in different roles and with differing interests and values are all provided with a flow of meaningful information, and where they have the opportunity for joint learning and innovative responses to this feedback from the environment and from other changes.
>
> It is this distributed intelligence which allows players in a community to anticipate and constructively address both individually and collectively the systemic problems the community continually faces and to deal with the threats and opportunities of natural and manmade disasters, the shifting global economy, and inequitable distributions of resources.
>
> **—Judith Innes[1]**

Now that we have offered a suggested framework for the creation and implementation of a comprehensive plan, let us look at a framework for monitoring our progress toward achieving goals and fulfilling a community's vision. We look at change and our perception of it, progress and how we monitor it, and outputs versus outcomes.

An important ecological principle to keep in mind in the creative process of both designing and carrying out a plan is diversity. Diversity was highlighted previously as an important feature of community input during the planning process and then again an important part of the plan itself. In monitoring the plan and its implementation and in gauging progress, a diversity of perspectives is important. This is not to say that clear indicators of progress cannot be defined; it is to say we must be certain that the process of defining indicators represents a diversity of perspectives.

Diversity is related to a certain amount of functional redundancy in the form of functional backups within a system. The more diverse a system is, the more stable it is, and the greater the number of built-in backups and thus resiliency. That is, when one part of the plan does not produce the desired result, another is available to perform a similar function and thus maintain the productive capacity of the system in question. Thus, the greater the diversity of input, the greater the probability that backup ideas will be part of the plan, thereby building resilience into the plan's implementation. Diversity suggests both flexibility and adaptability to change and naturally leads us to the question of how we generally respond to change, which in turn depends on how we perceive change.

Change and Our Perception of It

After designing a plan with a clear vision and clear goals, it is tempting to believe that, if we choose the correct measures of success, we will then be able to know if we have succeeded or failed. Here one must keep in mind that success and failure are not events but rather interpretations of events. If you and I both participate in the same event, for example, you may deem the event a smashing success, whereas I may interpret it as a dismal failure—and we are both correct from our own points of view. When, therefore, we have goals and defined ways for achieving the goals through the attainment of set objectives, it is indeed important to define benchmarks or indicators of progress, which we discuss in this chapter, but it is also important to know, short of using indicators and benchmarks, when we are off course.

Responding to the sense of being off course is best done in the collective, where inclusiveness and openness to the opinions and ideas of others is maintained and an indicator of progress is seen as a tool to help us reach a goal rather than becoming the goal itself. For example, a community that has worked together to create a comprehensive plan as well as indicators of progress might allow the process to languish by adopting the products—plan and indicators—as material representations of the desired future. Allowing a product to replace a process is also what happens when humans become too attached to a concept and convert it to something concrete; this is called *reification*. Shifting focus from community values—the intangibles—to measurable indicators of success could conceivably erode the values, which brings us to change, how we perceive it, and consequently how we respond to it.

How we plan for the future implies being aware of and finding a need to deal with change, which we all know is a universal constant of eternal novelty. Whether we are aware of it or not, we all change in response to change. It is inherent and is called self-organization.[2] Although a change may be minute and dysfunctional, we nevertheless respond to changes in our environment; in that, we have no choice. Being aware of the change, both external and internal, and seeking to understand the implications of the perceived change as well as the implications of our response result in a level of awareness that we believe contributes greatly to building and sustaining a sense of community.

Some change is so gradual that the period of time after which the change can be viewed as significant might be greater than a typical person's life span. Global climate change, for example, is still viewed by many, predominantly nonscientists, as a nonissue. In other words, there may not be the political will to push forward legislation that would result in adequate resources to increase our understanding of this matter. If, however, there was a consensus that enough evidence exists to merit the funneling of tax dollars into the appropriate research to obtain a better handle on this phenomenon, it would most likely happen, with or without overwhelming evidence that a crisis looms on the immediate horizon of the "invisible present" we mentioned in Chapter 1.

Now, it is necessary to revisit the invisible present, as Professor John Magnuson calls it, which is our inability to stand at a given point in time and see the small, seemingly innocuous effects of our actions as they accumulate over weeks, months, and years. Consider that all of us can sense change—the growing light at sunrise, the gathering wind before a thunderstorm, the changing seasons. Some people, living for a long time in one place, can see longer-term events and remember more or less snow last winter compared to the snows of other winters, that spring seemed early in coming last year, or that it is hotter than usual this summer.

In spite of such a gift, it is an unusual person who can sense, with any degree of precision, the changes that occur over the decades of his or her life. At this scale of time, we tend to think of the world in some sort of "steady state," and we typically underestimate the degree to which change has occurred. We are unable to directly sense slow changes, and we are even more limited in our abilities to interpret their relationships of cause and effect. This being the case, the subtle processes that act quietly and unobtrusively over decades reside cloaked in the invisible present.

The invisible present, writes Magnuson, is the scale of time within which our responsibilities for our planet are most evident. "Within this time scale, ecosystems change during our lifetimes and the lifetimes of our children and our grandchildren" (p. 495).[3] And, hidden within this scale of time are the interwoven threads of cumulative effects, lag periods, thresholds, and the various degrees of irreversibility that cause the practice of exploitive forestry to reverberate throughout the forested ecosystem.

Some liken the human experience of gradual change and the inability to sense the need to change to the phenomenon of placing a frog in cool water that is gradually brought to a boil. Apparently, a frog will not react to the imminent threat of being boiled alive because it does not detect the inevitable outcome of the gradually heating water and thus dies.

In 2000, I (Jane) spoke with Dr. Anthony Socci of the U.S. Global Climate Change Office in Washington, D.C. I raised the issue of the gradual nature of global climate change and the concern I have about the massive lack of attention to this matter, indicated, in part, by the absence of this issue from the platform of then-presidential candidate Al Gore, whose book *Earth in the Balance: Ecology and the Human Spirit*[4] addresses this issue at great length. His response was that *when* the level of carbon dioxide is four times greater than it was in 2000 (385 ppm),[5] or if we can see the imminence of that level being reached, then more attention will be given the problem. The current level is 395.55 ppm.[6] It is therefore necessary to urge communities and individuals to elevate their of consciousness to a higher level of awareness of cause and effect if they are to be prepared to respond positively, rather than react irrationally, to perceived threats.

Another example of the invisible present is the disappearance of species. In "The Sixth Extinction: How Large, How Soon, and Where?" a study by Stuart Pimm and Thomas Brooks, we learn that "humanity's impact has increased extinction rates throughout the world to levels rivaling the five mass extinctions of geologic history … [and] for a wide range of well-studied taxa, the known current extinction rates are several hundred times higher than the background extinction rate demonstrated by fossil records."[7] Based on our knowledge of the extinction rates of various species of birds, we can predict the potential future losses because they are directly linked to loss of habitat, which, according to the authors, "continues apace" and "in the next fifty years, deforestation will doom one half of the world's forest-bird species."[8]

In light of this information, one might ask what can be done to prevent future loss of habitats and thus species or, we add reservedly, the sixth extinction. What will it take for the global society to respond? How many species will be lost before significant actions are taken? Creating a significant decrease in carbon dioxide emissions—significant enough to reverse global warming, according to Anthony Socci—is unlikely. What is the threshold beyond which societies are moved to change or "self-organize" in response to change itself? Or, will human society wait until it is struck dead by changes of its own making? Other questions to ponder are the following:

1. How rapidly are we actually capable of understanding physical, cultural, or economic situations that nestle into a complex system that we do *not* fully understand?[9] Because this question is not asked, there is a massive illusion that we humans are somehow in control of nature, as evidenced by people, in awe of our having put a man on the moon, who now believe we are capable of solving all our global problems.
2. How able are our social institutions of perceiving the multidimensionality of issues and successfully communicating this information to the public?
3. How free is society to respond to this understanding, even if it can be achieved? Our sluggish response to continued population growth is due, in part, to strong opposition by some groups to equality for women and family planning.
4. How quickly can we self-organize or change in response to change?
5. How able are humans to perceive their own free will, and then, once perceived, under what conditions will they exercise it?

These questions are pivotal in looking at how we respond to changes in the world. We might also ponder what types of behavior characterize those humans who are best prepared to deal with this type of gradual change in an effective manner.

Russell Ackoff, in his 1974 book, *Redesigning the Future: A Systems Approach to Societal Problems*, describes four types of people and the ways each responds to the world.[10] Donella Meadows, in her article, "Chicken Little, Cassandra and the Real Wolf—So Many Ways to Think about the Future," suggests that one of the four types Ackoff describes—the "interactivist" is the most likely to bring about a sustainable society.[11] Interactivists, unlike Ackoff's inactivists, reactivists, and preactivists, are not willing to settle for the current state, return to the past, or get to the future ahead of everyone else. They want to design a desirable future and invent ways of bringing it about.

Interactivists work to prevent, not merely prepare for, threats and create, not merely exploit, opportunities. Interactivists struggle to achieve self-development, self-realization, self-control, and an increased ability to design their own destinies. They are not satisfiers or optimizers, but rather idealizers. To them, the formulation of ideals and visions is not by empty, imaginary exercises in utopianism but by necessary steps in setting the direction for conscious, sustainable development. Interactivists work to change the foundations, as well as the superstructure, of society, institutions, and organizations. They work to redirect the tide of change, not to resist, ride with, or ride ahead of it.

We agree with Meadows that such a way of approaching the future will help to create a sustainable society. The approach implies being conscious of the fact that we live daily in the intersection between yesterday and tomorrow, which means being aware of the present moment without losing track of the past and simultaneously inventing possible futures based on past events and current knowledge.

Consistent with the notion of optimal places for leveraging change within a system, the interactivists seek to make a *behavioral* change, to alter the very fabric of social consciousness. This type of behavior requires courage and self-knowledge. It requires fearlessness and deliberate living, in a way that accepts not having all the answers and honors the sacred space of ignorance in oneself and others. We can only teach other this way of being by risking living it ourselves.

What, one might ask, does this all mean in the context of having prepared a comprehensive plan with a vision, goals, and objectives to create programs and see the plan implemented through new policies, laws, and citizen initiative? It means continuing to live the identified values of one's community and not believe that any particular product, like a plan or indicators of progress, can replace the process of continually coming together in the conscious, deliberate creation of a sustainable community, or local living economy.

Thus, while models and paradigms, methodologies, and approaches will help us get where we want to go, it is only when each of us strives toward self-realization and development that the collective is most likely to succeed. It is when we can bring a fearless, peaceful, highly conscious view of ourselves and others to the world that we are most effective in our response to change. In so living, we will affect others, whether we see it or not, because this type of behavior not only is contagious but also proffers hope, a rare emotion in today's harried world. Lao Tzu, the sixth century BCE Chinese philosopher, said, "Leaders are best when people scarcely know they exist, not so good when people obey and acclaim them. When leaders fail to honor people; people do not honor them. But, of good leaders who talk little, when their work is done, their task fulfilled, the people will say, we did this ourselves."[12] Having said this, we must now face the question of creating measures of progress.

Creating Measures of Progress

There is a good body of literature on indicators of sustainability, sustainable forests, sustainable communities, and so on. Indicators, of course, are bits of information that highlight what is happening in the larger system. The purpose of creating indicators with which to measure progress is to show us how a system is working, whether it is going the way we want it to or not. Indicators tell us which direction a critical aspect of our community, economy, or environment is going in terms of our vision and goals.[13] Indicators are also educational in that those dealing with sustainable community should help communicate what sustainability means.[14]

Today, many in the United States look to the gross domestic product to determine how well we as a nation are doing economically. We discussed this previously and discussed the alternative notion of the genuine progress indicator.

Years ago, Robert Kennedy acknowledged the shortcomings of the gross national product, as it was then called, as an indicator of the nation's economic health:

> The Gross National Product includes air pollution and advertising for cigarettes, and ambulances to clear our highways of carnage. It counts special locks for our doors, and jails for people who break them. The Gross National Product includes the destruction of the redwoods and the death of Lake Superior. It grows with the production of napalm and missiles with nuclear warheads. ... And if the Gross National Product includes all this, there is much that it does not comprehend. It does not allow for the health of our families, the quality of their education, or the joy of their play. It is indifferent to the decency of our factories and the safety of streets alike. It does not include the beauty of our poetry or the strength of our marriage, the intelligence our public debate or the integrity of our public officials. ... The Gross National Product measures neither our wit nor our courage, neither our wisdom nor our learning, neither our compassion nor our devotion to country. It measures everything, in short, except that which makes life worthwhile; and it can tell us everything about America—except whether we are proud Americans.[15]

These sage words are a way of introducing how we go about designing indicators that tell us how well we are doing. Knowing how well we are doing in achieving the goals and objectives of a comprehensive plan can be determined in part from indicators and benchmarks, which are points of reference or standards against which measurements can be compared.[16] However, because the main feature of an indicator is its measurability, what do we do with the immeasurables? Can they be implied by what is measurable? Ideally, yes, if the indicators are based on critical issues, identified early in the planning process, which in turn are based on identified community values; in this case, the immeasurables will be "part and parcel" of each indicator. Indicators designed by a community can be very powerful and are intensely democratic since they are the community's own device for reporting to itself about itself.[17]

Most important, the community needs to remember the words of H. Thomas Johnson, professor of business administration at Portland State University in Oregon and distinguished consulting professor of sustainable business at Bainbridge Graduate Institute in Washington: "Perhaps what you measure is what you get. More likely, what you measure is all you'll get. What you don't (or can't) measure is lost."[18]

With this in mind, there are two general approaches to developing indicators for whole populations, such as a city or a county. One approach relies on technical experts to determine the appropriate indicator for measuring a trend. The other relies on a grassroots approach, and it is tied to participants' values. In determining indicators for a local living economy, we recommend an approach that combines technical advice and citizen values. In the Sustainable Seattle model, citizen values and needs drive the process, but scientific data and methods provide the foundation for indicators so that the selected metrics are understandable and valid. "The Sustainable Seattle process is iterative, with information moving among and being discussed by: (1) the general public, (2) a group of civic leaders, and (3) a group of technical advisors."[19]

Indicators are associated with each goal, objective, and strategy found in the municipality's comprehensive plan. The indicators provide data relating to progress that can be measured over time. Examples of sustainability indicators offered by Maureen Hart in her *Guide to Sustainability Indicators* are income disparity, water use per person, hours of work required to meet basic needs, percentage of frontline employees who attended employer-sponsored training, ratio of the number of hotel jobs to number of visitors, remaining farm acreage (number of hectares), energy use, and ecological footprint.[20]

Let us say, for example, that a critical issue defined during a community planning process is that the town is unable to handle the amount of tourism it is experiencing as revealed, in part, by the fact that visitors are forced to find lodging elsewhere during a typical summer season. The indicator, in this case, would be the number of motels and hotels (inns) or the number of rooms within a defined area. It is important to examine carefully, however, what constitutes a critical issue before converting it into an indicator.

In our example, it appears that the inability to accommodate the summer tourist trade is limited by the number of rooms available. Another response to this information might be to (1) make an effort to increase the tourist trade in what is currently seen as the off season, (2) accept the number of rooms now available and the scale of development this number of rooms represents, and (3) accept the loss in the lodging trade during the one season of the year. This, in fact, might even out the other effects of tourism on the city's infrastructure. Thus, an indicator of progress responding to this critical issue—the community's inability to provide adequate lodging during one season of the year—would be the number of additional off-season tourist attractions developed.

Suggested criteria for indicators offered by Sustainable Seattle are that an indicator is

1. Relevant
2. Easily understood by everyone, not just the experts
3. Compelling—illustrating community values
4. Desirable for use by the local media
5. Defensible logically and scientifically
6. Statistically measurable
7. Reliable, so that it can be measured consistently over time to produce comparable data
8. Appropriate in scale, suggesting the avoidance of composite indicators
9. Leading, or allowing a proactive rather than reactive response, or worse yet, crisis management
10. Policy relevant or of obvious relevance to potential policy formation[21]

In measuring our progress toward achieving the goals of a comprehensive plan, we can do at least three things, from the simple to the difficult: (1) monitor progress according to the timelines associated with specific objectives intended to carry out the plan and see if the city or municipality is on target; (2) develop indicators associated with critical issues, as described previously, which may or may not be tied directly to a definition of sustainability (if sustainable community indicators are preferred, the Maureen Hart guide is an excellent resource to guide this process); and (3) work toward monitoring the implementation of the vision and goals as well as indicators concurrently at both the government and community levels, which brings us to outputs versus outcomes.

Outputs versus Outcomes

The initiation and completion of objectives called for in the comprehensive plan are examples of outputs. They are generally quantitative in nature, for example, the enlargement of a city's community center. Outcomes are things that follow as a result or consequence.[22] They are more qualitative in nature than outputs and have to do with changed attitudes, higher levels of knowledge in the community, or altered human behaviors. Outcomes are, of course, more difficult to measure. For example, one reason for expanding the community center was to provide space for more community gatherings and to help build a sense of place. How do we know that sense of place, an outcome, has resulted from expanding the community center—an output? An indicator could be devised to measure this, or a study could be done.

Maureen Hart offers a number of indicators for "connectedness," which is an aspect of a sense of place. These include the "number of community gardens, percent of people who say they have enough, average distance between residences of extended family members, citizen participation in community projects, neighborhood livability as rated by residents, neighborliness, percent who say most people can be trusted."[23]

A study done in 1995 for Parks Canada sought to establish a protocol for defining the human concept of place in the Lake Superior Basin.[24] The study was conducted through interviews with basin residents. First, the boundaries of each person's "place" was determined; then, it determined what sorts of things connected the person to his or her place—was it the landscape, family connections, type of lifestyle, work-related or social interactions? Then, the interview sought to find what deep meanings this sense of place held for the person, especially if there was some spiritual aspect to the connection. Finally, the research attempted to determine whether the person could stretch these "deeper meanings" over a larger area. Thus, it is possible to obtain some idea at least of existing intangibles, like a sense of place, as well as attitudes and other feelings.

Ultimately, progress can be measured and felt in many ways. Some people feel that true progress will have happened when poverty and prosperity are detached from the measurable things of a material nature, as so eloquently stated by Robert Kennedy with respect to what the gross national product does *not* measure—courage, wisdom, learning, compassion, service to others, and so on.[25] It will be a grand day indeed when we can call ourselves the wealthiest nation in the world and look to the health of our children without having to hide the poverty that any American child lives in, the high level of volunteerism and peace in the world, or a lowest level of alcoholism and its violence in the world.

So, the beginning of the planning process, when a community determines its core values, is the best time to describe what indicates the presence of that which we value in our community and work toward keeping those features alive and well as our adult legacy to those who follow.

Summary

Chapter 9 is a comprehensive look at how to monitor the progress of a land use plan once it is implemented. Chapter 10 discusses how to maintain community engagement.

Endnotes

1. Judith Innes, coauthor of *Planning with Complexity: An Introduction to Collaborative Rationality for Public Policy*. Routledge, Oxford, England, 2010.
2. Donella Meadows. "Places to Intervene in a System." *Whole Earth* Winter (1997). http://center.sustainability.duke.edu/sites/default/files/documents/system_intervention.pdf (accessed March 9, 2013).
3. John J. Magnuson. "Long-Term Ecological Research and the Invisible Present." *BioScience* 40 (1990):495–501.
4. Senator Al Gore. *Earth in the Balance: Ecology and the Human Spirit*. Houghton-Mifflin, New York, 1992.
5. James Hansen, Makiko Sato, Pushker Kharecha, and others. "Target Atmospheric CO_2." *Open Atmospheric Science Journal* 2 (2008):217–231. http://www.columbia.edu/~jeh1/2008/TargetCO2_20080407.pdf (accessed March 5, 2013).
6. CO_2 Now. Home page, "What the World Needs to Watch." http://co2now.org/ (accessed March 5, 2013).
7. Stuart L. Pimm and Thomas M. Brooks. "The Sixth Extinction: How Large, How Soon and Where?" In: *BioDiversity*, P. H. Raven and T. Williams (eds.), 46–61. Washington, DC: National Academy Press, 1999.
8. Ibid.
9. Braden R. Allenby. *Industrial Ecology: Policy Framework and Implementation*. Prentice Hall, Upper Saddle River, NJ, 1999.
10. Russell Lincoln Ackoff. *Redesigning the Future: A Systems Approach to Societal Problems*. Wiley, New York, 1974.
11. Donella Meadows. "Chicken Little, Cassandra and the Real Wolf—So Many Ways to Think about the Future." *Whole Earth* 96 (1999):106–111.
12. John Heider. *The Tao of Leadership*. Bantam Books, New York, 1986.
13. Sustainable Seattle. "A Primer for Creating New Measurements of Progress." Seattle, WA: Sustainable Seattle, 1996.
14. Maureen Hart. *Guide to Sustainable Community Indicators* (2nd ed.). Hart Environmental Data, Andover, MN, 1999.

15. Sustainable Seattle. "First Set of Indicators." 1993. http://www.sustainableseattle.org/programs/regional-indicators/133-first-set-of-indicators (accessed March 9, 2013).
16. Sustainable Seattle, "A Primer."
17. Ibid.
18. James Considine. "What You Measure Is What You Get?" December 22, 2008. http://www.isixsigma.com/community/blogs/what-you-measure-what-you-get/ (accessed March 3, 2013).
19. Sustainable Seattle. "Fourth Set of Indicators." n.d. http://www.sustainableseattle.org/programs/regional-indicators/124 (accessed March 5, 2013).
20. Hart, *Guide.*
21. Sustainable Seattle, "A Primer."
22. The following discussion in this paragraph is based on American Institute of Architects. *Designing Your Town*. 199
23. Hart, *Guide.*
24. The following discussion in this paragraph is based on American Institute of Architects, *Designing*.
25. Sustainable Seattle, "First Set."

10

Ongoing Community Engagement: Citizens as Planners

> When we see land as a community to which we belong, we may begin to use it with love and respect.[1]
>
> **—Aldo Leopold**

Throughout this book, we have suggested the need for a major shift in values and beliefs. There will be many, no doubt, who seriously question, if not simply refuse to accept, that such a shift is even possible, that we can alter the course we are on—one toward dangerous societal hardships and severe planetary depletion of the biophysical services we humans require for our very existence. Recognizing that many are perplexed, if not frightened, by the current conflict between human desires and the biophysical necessities of the planet, we have offered a historical context, as well as theoretical perspectives on human behavior, with the hope of removing barriers to creative problem solving. While this knowledge alone may not initiate profound creativity, perhaps trusting in a process that is open, transparent, and inclusive, as well as ongoing, will at least help those who choose to engage in planning for the future of their community release judgment and fear. The paradox of letting go of fear and judgment (control) and, by so doing, achieving a greater influence over one's future (control) could be a highly rewarding, even transformative, experience.

The purpose of this chapter is to examine ways of keeping the community engaged in a planning process that will forge a new way of being in community. We suggest this could be a return to something for which we have an inherent ability and a natural affinity, that deep inside of all of us, we know what is right for Earth and the well-being of all life. Deep inside all of us there is a longing for intimacy—being a part of a greater whole, whether community, family, or even the web of life. We might call this the ecological unconscious, discussed previously. And, we ask, could it be that the ecological unconscious is in a constant struggle with fear—some conscious, some unconscious—that works to sever connections with the whole or simply feeds disassociation with the whole?

While history tells us that human behavior has always had a dark side and that destruction of life is nothing new (based on the extinction of species, the demise of certain cultures, and other permanent losses), we believe it is always possible for profound change to take place in ways that we have not yet imagined. In other words, whereas much of the longing for change seems to represent values associated with a return to the simpler lifestyles of more "primitive societies" as humanity's only hope for survival with any kind of dignity and well-being, we think it best to remain open to all possibilities.

To remain open to the possibilities, we best begin establishing new ways of being in community, ways that retain and celebrate the best features of community as it now is, while remaining open to the need for course corrections; learning how to assess what works and what does not; and doing whatever we can to stimulate and support creativity in ourselves and others, all the while being comfortable with not having all the answers, remaining at ease with uncertainty. By remaining open to new possibilities, we can deploy our ability to transform obstacles into opportunities for the imagination to seize and begin to see obstacles as guests to welcome rather than black helicopters in the neighborhood.[2] We can even learn how to "dance" with the obstacles rather than wishing to crush them, knowing that they most likely have valuable information to convey to us about us and our circumstances.

To be open to new possibilities for a better future, however, means engaging in opposing points of view. It means finding and allowing the artist within to come out and play and embrace the prickliness of complexity and risk with a sense of wonder. It also means simply starting to talk to one another again, as counseled by Peter Senge, in a 1999 keynote to a conference on systems thinking and action.[3]

To "grow" a new way of being in community as a result of a planning process, we must understand our audience and the preferred ways all segments of a community send and receive messages. Because this is such an important part of the planning context, we must also be certain the message itself is clear. Part of solidifying the planning context and retaining community engagement is a clear understanding of the problems as well as the knowledge of what is working elsewhere that might apply to local problems, along with promising global and societal trends. For example, what are some of the indicators that a societal "paradigm shift" may be occurring that will necessitate the merger of human desires with the biophysical constraints of social-environmental sustainability? In addition, what are the barriers to the emergence of a new society?

Knowing Our Audience

Today, the generations are separated in part by the ways they prefer to communicate—both receiving and sending messages. Studying local demographics, particularly age cohorts, will provide a good start in determining the best means for distributing information.

These days, the "next big thing" is usually about technology-based communication, for which revolutionary changes have a significant role in characterizing the demographic cohorts. Generation Y (born between 1980 and 2000), for example, grew up with technology and rely on it not only to perform their jobs better but also to interact with friends and family. Armed with laptops, tablets, smartphones, androids, and other mobile technologies, members of this generation are plugged in 24 hours a day, 7 days a week.

These are the true digital natives who prefer using electronic means to communicate as opposed to face-to-face meetings. This generation is far more fluent and comfortable with technology than any other group, preferring to communicate through e-mail and text messaging rather than face-to-face contact and, in the realm of business and education, preferring webinars and online technology over traditional lecture-based presentations.[4]

Generation X, on the other hand (born 1966–1979), tends to use technologies that support their lifestyles, such as online shopping and online banking. Their lives are hectic with careers and families, and they embrace the digital convenience. While the number of Gen Xs reading blogs and using social media is rising, they still fall behind Gen Y.[5]

Although advances in communication technology are nearly constant and to some daunting, we need to pay attention and stay current if our community planning processes are to remain appealing, attractive, and durable. In doing so, we have a better chance of successfully engaging a diverse audience by communicating in ways that we will most likely be heard and comprehended. Staying current with the predominant forms of communication in our communities will help us stay relevant, as resistant as some might be to the notion that human face-to-face contact is being replaced with technology-based human connection.

What Do We Need to Be Communicating and How?

Simply stated, the message is that we need to discover—or rediscover—our connection to the rest of life on the planet and know that nurturing it will improve our chances for survival while contributing significantly to our quality of life. In this book, we suggest that, as we plan for what happens to ourselves and one another—and consequently to the land of which we humans are an inseparable part—we can begin to see, understand, and feel our connection to Earth, which again echoes Aldo Leopold's words, "When we see land as a community to which we belong, we may begin to use it with respect and love."[6]

Our message is motivated by perceived risk. That is, we are saying we humans are at risk, given the way we are currently operating in terms of land use practices. Our practices need to change. This reality is strengthened by the recently introduced concept of the Anthropocene to denote the current interval of anthropogenic, global environmental change. The term was coined by ecologist Eugene F. Stoermer and widely popularized by the Nobel Prize-winning atmospheric chemist Paul Crutzen. The Anthropocene is an informal geologic chronological term serving to mark the evidence and extent of human activities that have had a significant global impact on Earth's ecosystems. Crutzen argues, "The influence of human behavior on the Earth's atmosphere in recent centuries as so significant as to constitute a new geological epoch for its lithosphere."[7]

Based on mounting evidence and concern, how can we communicate our message in ways that produce or encourage responsible action? Previously, we discussed engaging the community in a planning process by emphasizing that change is imminent, and that these changes are occurring in the world, in one's region, and particularly in one's own community. And, further, every community can have some "say" over how such change occurs.

To reiterate, successfully conveying this message means not only determining the right language but also the right medium. It is incumbent on those leading the planning process to tap into as many forms of communication as there are identified demographic groupings in the community. Social media are a growing source of news and must be considered a part of any communication strategy. This will likely mean creating a Facebook page (1 billion users in 2012) for the planning initiative, tweeting, text blasts, and YouTube videos (perhaps of such things as educational slidecasts), and so on.[8]

We can expect a full range of response to this message, as with any message proffered to a community at large. It may spark interest; it may not. In the realm of public meetings, we often see an attempt to "hook" others into involvement through fear, and while fear does indeed draw people into the public arena, as a motivator it seldom works for building long-term solutions or love of community and sense of place.

We also know that when people become aware of a risk or a threat, they are inclined to become engaged in fear of the unknown and thus want to control circumstances relative to their own well-being. Conflicts that surface from the fear of risk are often polarized between those who trust scientific analysis and believe the demonstrable data and the scientific interpretation based on that data and the general public, who neither trust nor accept the data or its interpretation—a modernist-traditionalist conflict discussed further in this chapter. Understanding this common dynamic is important when seeking to engage the public in meaningful ways.

To engage a community in planning for its future, it is important for those leading the planning process to (1) make sure the community recognizes and accepts agency responsibility for making decisions associated with the planning process; (2) state the reasons for planning; (3) describe the planning process and how an individual can become involved; (4) describe what the benefits will be to the community; (5) describe the possible scenarios should no planning be undertaken; (6) provide further information to help citizens evaluate the usefulness of this work; (7) pretest messages; (8) evaluate your efforts; (9) know your audience or audiences (do not underestimate their intelligence or overestimate their knowledge); (10) deal directly with uncertainty; (11) demonstrate the collaborative nature of the endeavor; (12) respond promptly to comments and concerns; and (13) publicize both the meetings and the accomplishments. The last two actions are forms of recognition that function to draw citizens into the process.

In a community just outside Santa Cruz, California, in a neighborhood called Live Oak, county planners conducted a planning effort in 1992 designed to establish a direction for land use, design, and traffic circulation. The neighborhood was initially engaged through a series of community meetings, where residents identified what they felt needed protection and what needed to change.[9] Here, we offer a critical suggestion: Allow participants to write their feelings down while at home or somewhere besides a social gathering to accommodate the introverts in the audience. This is an important step—one that is virtually always overlooked. To better understand the vital importance of this compassionate gesture, we recommend reading Susan Cain's insightful book, *Quiet*.[10]

Following the meetings in Live Oak, a workbook was prepared for the community that presented (1) the ideas gathered at the preliminary meetings; (2) an expression of the citizen's ideas in alternative scenarios for land use, transportation, and parks and recreation for 10 special areas in the community; and (3) a section soliciting further ideas for design and placement criteria for new housing in the community.[11] The community was thus immediately engaged by being asked about neighborhood concerns and requirements, which then formed the foundation for a workbook that was used in the planning process. The workbook recognized the willingness of the residents to participate and express themselves. It also made the ideas of the community more tangible while being a useful tool in the planning process.

The process undertaken in Santa Cruz County represents sensitivity to community and its sense of place. This is one characteristic of a powerful communication strategy. Similarly powerful in terms of keeping participants engaged and inspired is informing them of emerging and promising national, if not global, trends toward sustainability.

Is a "Paradigm Shift" Occurring?

At the societal level, some believe that a "bottoming out" may be under way. Reaching the bottom, whether personally or socially, frequently results in transformative change. Are we at the point of a major paradigm shift in how we view the world? Is a social transformation occurring?

The "new normal," which may be contributing to a "new low," is the growth in human interaction through virtual communication, leading to a more depersonalized society. Virtual communication is more often written and lacks emotions and nuances that are evident in person-to-person contact, where tone of voice, physical motions, hand gestures, speech patterns, and facial expressions create a more humanizing experience through the unspoken part of language, which can lead to clearer interpersonal understanding.

Virtual communication also feeds a growing sense of independence and the perceived lack of *need* for human contact. We know, for example, that we can conduct a business, connect with friends and family, do much of our shopping, research our medical questions, keep track of finances, pay our taxes, and vastly more without having to leave our homes. Computer technology, as well as automated services, which take many forms from automated bank tellers to robotic surgery, can for the most part decrease, if not eliminate, our need for direct human contact for many of our material needs. In addition, rapid access to information, as well as an enhanced ability to manipulate the information (but without time to assimilate it[12]), has fostered a heightened illusion of being in control of how the world works. As our sense of perceived control expands, our awareness of impending system failures declines, and our contact with other humans decreases accordingly—making us extremely vulnerable. And, if this vulnerability and resulting fragility were recognized, could it lead to adaptation, self-organization, and perhaps a new era?

Although the modern age is characterized by individualism, materialism, and the illusion of rationality, the new age or age of sustainability will be characterized by community, spirituality, and intuition, as suggested by author Dorothy Lagerroos 17 years ago.[13] Are the values of mobility, no limits, growth, and anthropocentrism now being replaced with the values of rootedness or sense of place, the acceptance of limits, the notion of growing simplicity (less is more or when is enough, enough), and humans as an inseparable part of nature?

Others, like author and sociologist S. J. Goerner, believe a mass movement is afoot. In an article that discusses how scientific and social movements are meeting in a quest to remake the world, Goerner argues that a "great turning is going on."[14] The great turning is difficult to perceive because "in a specialist world, no one has time for integration." Likewise, David Korten sees a great turning and makes the following case:

> Humans are a choice-making species that at this defining moment faces both the opportunity and the imperative to choose our future as a conscious, collective act. We can no longer deny the need nor delay our response. A mounting perfect economic storm is fast approaching. A convergence of climate change, peak oil, and the financial instability inherent in an unbalanced global trading system will bring an unraveling of the corporate-led global economy and a dramatic restructuring of every aspect of modern life.[15]

But, we know that some major trends go unnoticed. One such trend, recognized in the writings of Richard Florida, referenced previously (2012), is based on a 1996 study of the U.S. population by sociologist Paul Ray, in which he found that a quarter of the U.S. population fits into a group he calls Cultural Creatives. "Until ten years ago," said Ray, "the U.S. Population was composed of two main cultural groups, Modernists and Traditionalists. … Modernists ground themselves in science, technology, and industry. They tend to disdain older traditions. Traditionalists find their solidity in church and community and have doubts about the course that Modernists set."[16]

By 2000, Ray and coauthor Sherry Ruth Anderson claimed there were 50 million American Cultural Creatives with nearly 80–90 million European Cultural Creatives. In their 2001 book, *Cultural Creatives: How 50 Million People Are Changing the World,* Ray and Anderson describe the key characteristics of Cultural Creatives, which include a love of nature, a deep caring about its preservation, and its natural balance; a strong awareness of the planet-wide issues (such as climate change, poverty, and so on); a willingness to pay higher taxes or spend more money for goods if that money goes to improving the environment; a heavy emphasis on the importance of developing and maintaining relationships; an intense interest in spiritual and psychological development; and recognition of spirituality as an important aspect of life but at the same time worry about religious fundamentalism.[17]

Cultural Creatives, most likely without knowing it, are taking counsel from Sitting Bull, the Sioux chief, who told his warriors, after their victory at the battle of the Little Big Horn, not to rejoice. He said, in essence, that the white man was not going to go away, that the Indian peoples must choose the best of their old ways and the best of the white ways and put them together for that was their path into the future. Likewise, Cultural Creatives must choose the best of the traditionalists' ways and the best of the modernists' ways and put them together as they seek a new society that integrates science, technology, intuition, spirituality, and community for the health and well-being of the entire Earth.

The new path sought by Cultural Creatives is in response to the deterioration of systems, including education, economics, and politics, which were developed and guided by a way of thinking, popularly known as the machine age. Today, Cultural Creatives understand that the world is like a spider's web: One cannot touch any single part of it without sending vibrations throughout the entire web complex. We are now in what some call the information or digital age, in which global communications and networking are shaping modern society and mirror in many ways the web of connections that define nature.

Although scientists have known of such interconnectedness for some time, especially physicists, natural historians, and ecologists, it is becoming increasingly apparent that the mechanical notion of how the world works is rapidly falling into permanent disrepair because scientists finally have the tools with which not only to explore how interdependent systems work but also to understand what such interdependence really means. "The result [of such knowledge]," says Ray, "is a sudden jolt. We have grossly underestimated the role interdependence plays in shaping everything."[18]

Thus, while the many and various applications of new technology may separate us from one another in the personal sense, this very technology may also be the linchpin securing our understanding of the interrelatedness of the world as a whole. It may be quite a race, however, between the forces of separation and those of integration. Nevertheless, even if science is able to more fully comprehend interdependence, the effective communication of this information to the global populace will be crucial to the race being won by integration and the emergence of what Goerner calls the "integral society."

Barriers to Overcome

While there seems to be growing evidence of the emergence of an integral society, a great turning, or an age of sustainability, what are the barriers that we can expect that attempt to thwart full emergence? This is not an idle question because there are always barriers to overcome in any new endeavor; they are simply part of the price of meaningful change. Author Daniel J. Boorstin says it best:

> The obstacles to discovery—the illusions of knowledge—are also part of our story. Only against the forgotten backdrop of the received common sense and myths of their time [and ours] can we begin to sense the courage, the rashness, the heroic and imaginative thrusts of the great discoverers. They had to battle against the current "facts" and dogmas of the learned.[19]

Government structure, which generally precludes authentic community involvement, is certainly a barrier. But, examples of changing government structure are starting to surface.

The U.S. Forest Service is certainly out in front of other federal agencies in bringing new language and new concepts to the formation realm of policy and law. In 1993, the Northeastern Area of the U.S. Forest Service published *An Ecosystem Approach to Urban and Community Forestry,* which discusses the tenets of deep ecology as a way of guiding a needed shift from an "industrio-scientific paradigm" to a "bioregional paradigm."[20] And, in 1998, the U.S. Forest Service began an initiative to develop comprehensive monitoring programs to ensure sustainable land management. It worked with six national forests that volunteered to be "pilot forests" in a project aimed at developing criteria and indicators of sustainability, which, when implemented, "provide measures of ecological, social and economic well-being."[21]

While the U.S. Forest Service had been edging toward such innovations long before 1998, the threats to loggers and other entities with a financial stake in what happens to the nation's forests started to gain attention. In Minnesota, for example, a group of independent loggers sued the "Superior Wilderness Action Network" because their comments on timber sales by the U.S. Forest Service delayed by several months the awarding of timber-harvesting permits.[22]

The plaintiffs claimed that the U.S. Forest Service is being influenced by the notion of deep ecology, and that both the Superior Wilderness Action Network and the New Mexico-based Forest Guardians are pushing this "religion" onto the service. The attorney for the plaintiffs, Stephen Young, stated that, "If deep ecology bites into the way the Forest Service, the National Park Service, the Department of Natural Resources, and counties manage, then our natural surroundings will be governed by the religious point of view." Young went on to say that, "In a democratic society, the secular point of view is what should prevail."[23]

So it is that our greatest challenge in fostering a shift in the worldview or a social transformation will be the fear of loss—most often in the form of lost profits, as represented in this story—and possibly by the fear that somehow separation of church and state is compromised when a spiritual dimension is brought into government policy and law.

"What is going on at any point in time is not a matter of right versus wrong or good versus evil. It is simply a reflection of where the mainstream of consciousness is at this time," states Frederick J. Deneke, assistant director of the U.S. Department of Agriculture Forest Service Cooperative Forestry.[24] And, it is precisely this perspective that we believe will disarm fear. It will enable us to integrate our minds with our hearts, work collaboratively to find "practical and peaceful solutions and move forward together to affect the land in a way that benefits all."[25]

In addition to the wariness around the notion of spirituality seeping into government policy, as demonstrated by those opposing the U.S. Forest Service's incorporation of elements of deep ecology into Forest Service policy, there is the matter of the U.S. political system. Barriers to the coming of age of a society built on the principles and concepts of sustainability include the U.S. political system.

Some people, including Princeton professor of economics and international affairs Paul Krugman, believe the U.S. political system is in disarray. He sees this as a structural problem wherein the system "has been warped and paralyzed by the power of a small, wealthy minority." Today, in Washington, D.C., Krugman notes there is "a combination of bitter partisanship and intellectual confusion—and both are … largely the result of extreme income inequality … [and] we have responded to the crisis with a mix of paralysis and confusion—both of which have a lot to do with the distorting effects of great wealth on our society."[26] Such an interpretation of our current struggles departs from the somewhat tired right versus left paradigm.

Accordingly, the 2008 book by Sally Goerners, Robert G. Dyck, and Dorothy Lagerroos, *The New Science of Sustainability: Building a Foundation for Great Change*, discredits the right-left dichotomy[27] and rather sees the real dichotomy as between an oligarchy and a partnership democracy[28]—"oligarchy" being a synonym for rule by the rich (Aristotle).[29] "Partnership democracy" rests on the concept of *partnership*, a term used throughout this book in reference to a "more cooperative, inter-connected society we all see emerging, although still under the radar," according to coauthor Dorothy Lagerroos, who goes on to note that "'partnership democracy' is not a specialized term, just a general one to stand against the dominator society we see burning itself out."[30]

Engendering partnership democracy, as part of our local living economies, will require moving away from oligarchic social structures to integrated systems-based, agile and dynamic, interlinked structures, just as we know exist in nature. And, as in a healthy ecosystem, the "wealth" is shared, as alluded to in the model for community planning we propose in which each "node"—human capital, natural capital, and financial capital—both feeds and is fed by others. Furthermore, to create local living economies, we suggest following a process that models the change necessary to lead to a higher level of consciousness and thereby adapt our lives, our political structures, our economies, and our communities toward social-environmental sustainability so that we humans, as well as our fellow travelers and the biophysical systems on which we all depend, might thrive on this planet. Last, in overcoming barriers to social transformation, we urge those leading communities to bring artists and children into the planning process as we move into an era where the nexus between science, art, and religion will be understood and will represent an age of sustainability—perhaps a new version of the "melting pot."

Artists as "truth tellers"[31] have much to contribute to planning for a more sustainable future—as do the hearts and minds of children who will inherit the wisdom or folly we adults bequeath them with the outcome of our decisions.

Mrill Ingram's art–science collaborations in the realm of environmental remediation convey the value of bringing scientists, artists, and the community together to solve ecological challenges, take action in the context of social complexity, and in so doing, further environmental science. Ingram references the work of artist Frances Whitehead of the School of the Art Institute of Chicago, who teamed with the city of Chicago and students at Chicago State University to plant a bioremediation project at an abandoned filling station lot. Whitehead lauds the artist's way of knowing that provides a "lateral heuristic methodology," a horizontal process of thinking that is critical to taking steps toward improved sustainability. "With an artist as lead agent, these projects capitalize on human memory and feeling while at the same time attending to environmental degradation, offering a refreshing set of ideas about sustainable development in the Anthropocene" (p. 10).[32]

Giving a role to artists in our community planning processes will engender an appreciation for the "artist" within all who participate and thus provide another means of igniting the creativity so critical to problem solving in the world. The resulting empowerment could quite possibly yield other self-defining experiences and greater knowledge of self—the "engine of our spiritual quest and the real source of power in our lives."[33]

Summary

In sum, one of the purposes of writing this book was to help people understand that the decisions we make today for our communities sometimes become the problems our children and their children will inherit tomorrow.

It is imperative, therefore, that people become aware of the long-term effects of their decisions because children are all too often one of the two silent parties in all social-environmental decisions—the land and its productive capacity is the other. All parties must understand the environmental and economic circumstances to which they are committing the future because if the outcome is destructive and thus a deficit in terms of either the children's future options or the ecosystem's productive capacity, it is analogous to "taxation without representation," and that goes against everything for which a democracy stands.

Once the people of a community have defined themselves culturally, they then decide what legacy they want to leave their children. Then, and only then, are they ready to craft a community vision and land use plan because only now do they really know what they want.

Endnotes

1. Aldo Leopold. http://www.goodreads.com/author/quotes/43828.Aldo_Leopold?page=1 (accessed March 21, 2013).
2. For inspiration, read Rumi's poem, "The Guest House." http://www.panhala.net/archive/the_guest_house. http://www.gratefulness.org/poetry/guest_house.htm (accessed March 17, 2013).
3. Peter Senge. Keynote speech. Systems Thinking and Actions Conference. Boston, November 5, 1999.
4. The foregoing discussion is based on Sally Kane. "Generation Y." n.d. http://legalcareers.about.com/od/practicetips/a/GenerationY.htm (accessed March 13, 2013).
5. The foregoing discussion is based on Helen Leggatt. "What's the Difference between Gen X and Gen Y?" July 24, 2008. http://www.bizreport.com/2008/07/whats_the_difference_between_gen_x_and_gen_y.html# (accessed March 13, 2013).

6. Leopold quotation.
7. Foregoing discussion is based on (1) Jan Zalasiewicz, Mark Williams, Alan Smith, and others. "Are We Now Living in the Anthropocene?" *GSA Today* February 2008. http://www.geosociety.org/gsatoday/archive/18/2/pdf/i1052-5173-18-2-4.pdf (accessed March 20, 2013); and (2) "Anthropocene." http://en.wikipedia.org/wiki/Anthropocene (accessed March 20, 2013).
8. Anthony Curtis. "The Brief History of Social Media." 2013. http://www.uncp.edu/home/acurtis/NewMedia/SocialMedia/SocialMediaHistory.html (accessed March 16, 2013).
9. Brady and Associates, in association with Korve Engineering. *Community Plan Workbook for Live Oak*. Santa Cruz County, CA: Brady and Associates,, 1992.
10. Susan Cain. *Quiet*. Broadway Paperbacks, New York, 2013.
11. Brady and Associates, Korve Engineering, *Community Plan*.
12. For more discussion of the impact of the Internet on our brains, see a review of *The Shallows*: Michael Agger. "The Internet Diet." June 7, 2010. http://www.slate.com/articles/arts/books/2010/06/the_internet_diet.2.html (accessed March 18, 2013).
13. Dorothy Lagerroos. "Conflicting Worldview?" Classroom presentation. Northland College, Ashland, WI, 1996.
14. Sally J. Goerner. *After the Clockwork Universe: The Emerging Science and Culture of Integral Society*. Foris Books, Edinburgh, UK, 1999.
15. David Korten. *Great Turning*. http://www.davidkorten.org/great-turning-book (accessed March 19, 2013).
16. Paul Ray. "The Rise of Integral Culture." *Noetic Sciences Review* 37 (Spring 1996).
17. The foregoing discussion is based on "The Cultural Creatives." http://en.wikipedia.org/wiki/The_Cultural_Creatives (accessed March 18, 2013).
18. Ray, "The Rise of Integral Culture."
19. Daniel J. Boorstin. *The Discoverers*. Vintage Books, New York, 1983.
20. USDA Forest Service, Northeastern Area. *An Ecosystem Approach to Urban and Community Forestry: A Resource Guide*. 1993.
21. Phyllis Green, forest supervisor, Ottawa National Forest. Invitation to stakeholders to participate in indicators project, 1999.
22. Dennis Lien. "Suite: Religion, Logging Don't Mix." *St. Paul Pioneer Press* (St. Paul, MN) October 2, 1999.
23. Ibid.
24. Frederick J. Deneke. "Forestry: An Evolution of Consciousness." *Journal of Forestry* 96 (1998):56.
25. Ibid.
26. The foregoing discussion is based on Paul Krugman. "Plutocracy, Paralysis, Perplexity." *The New York Times* May 4, 2012.
27. Sally Goerner, Robert G. Dyck, and Dorothy Lagerroos. *The New Science of Sustainability: Building a Foundation for Great Change*. New Society, Gabriola Island, BC, Canada, 2008.
28. Comment based on review of *The New Science of Sustainability: Building a Foundation for Great Change*. http://www.amazon.com/The-New-Science-Sustainability-Foundation/dp/0979868319/ref=sr_1_cc_1?s=aps&ie=UTF8&qid=1363547859&sr=1-1-catcorr&keywords=Sally+Goerner (accessed March 17, 2013).

29. Jeffrey A. Winters. "Oligarchy and Democracy." *The American Interest* 7 (November/December 2011).
30. Dorothy Lagerroos. Personal correspondence with Jane Silberstein via e-mail, March 22, 2013.
31. David Korten. *Agenda for a New Economy: From Phantom Wealth to Real Wealth* (2nd ed.). *Berrett-Kohler*, San Francisco, 2010.
32. The foregoing discussion is based on Mrill Ingram. "Sculpting Solutions: Art–Science Collaborations in Sustainability." *Environment: Science and Policy for Sustainable Development* 54 (2012):24–34.
33. Richard Thieme. "Between Transition." November 30, 1999. http://www.thiemeworks.com/between-transitions/ (accessed March 17, 2013).

Index

U

For Product Safety Concerns and Information please contact our EU
representative GPSR@taylorandfrancis.com
Taylor & Francis Verlag GmbH, Kaufingerstraße 24, 80331 München, Germany

www.ingramcontent.com/pod-product-compliance
Ingram Content Group UK Ltd.
Pitfield, Milton Keynes, MK11 3LW, UK
UKHW022015180926
475701UK00003B/367
9780367868048